Global Security and Intelligence Studies

Also from Westphalia Press
westphaliapress.org

Volume 7, Number 1 • Summer 2022

Carter Matherly , Matthew Loux & Jim Burch, editors

Westphalia Press
An imprint of Policy Studies Organization

Global Security and Intelligence Studies
Volume 7, Number 1 • Summer 2022

Westphalia Press
An imprint of Policy Studies Organization
1367 Connecticut Avenue NW
Washington, D.C. 20036
info@ipsonet.org

ISBN: 978-1-63723-817-2

Interior design by Jeffrey Barnes
jbarnesbook.design

Daniel Gutierrez-Sandoval, Executive Director
PSO and Westphalia Press

Updated material and comments on this edition
can be found at the Westphalia Press website:
www.westphaliapress.org

Global Security and Intelligence Studies
Volume 7, Number 1 • Summer 2022

(cont'd.)

Policy Relevant Commentary

Editorial Welcome

Welcome to the first issue of our Seventh edition! *Global Security and Intelligence Studies* is a double-blind peer-reviewed academic journal that aims to bridge the two-way gap between academia and practitioners. We serve as common ground to a diverse and growing audience ranging from policymakers to academics to operators on the front lines. *GSIS* strives to provide work pertaining to the most current and relevant topics in an ever-evolving and rapidly expanding threat-scape.

The summer issue is packed with informative and instructive research that spans the psychological, cyber, political, and strategic domains. We are excited to once again open our edition with a Graduate Lectern. Andrea Mosila presents an illuminating discussion on Gorbachev's psychological traits as a pathway for a peaceful end to the Cold War in "*Mikhail Gorbachev: A Transformational Leader*".

The term 'integration' is a key concept in today's intelligence community. The concept was represented 51 times across the 2019 national Intelligence Strategy. David Kritz explores how cognitive diversity impacts the intelligence community's ability to integrate across its 18 agencies in "*Coming Together: Strengthening the Intelligence Community Through Cognitive Diversity*". Casey Skvorc and Nicole Drumhiller continue the application of psychological analysis for strategic purposes by turning the lense of environmental development on Syrian President Bashar al-Assad. Their article "*Physician, then Political Dictator: Bashar al-Assad, President of the Syrian Arab Republic*" explores Al-Assad's life and influences to identify key attributes which have led to his responsibility for the death or disappearance of over 500,000 people in Syria's prolonged civil war.

In 1981 The U.S. General Accounting Office issued their seven page account on "The Government Brain Drain" detailing failures that had led to a drain on top talent and the impacts of such a drain. Ryan Burke and Jahara Matisek explore the strategic implications of this reality as a tool against adversarial nations through the lense of a great power competition and clashing political ideologies. Thier research "*A Soft Path to American Hegemony in the 21st Century: A U.S. Strategic Brain Drain Policy against Great Power Competitors*" explores the soft-power principals available to influence strategic competition in the 21st century. Continuing the Great Power Competition discussion Jonathan Schillo presents research addressing how can the United States effectively align military and diplomatic efforts

doi: 10.18278/gsis.7.1.1

to deter the People's Republic of China expansionism in the Indo-Pacific Command region in his article "*Politics With Other Means Aligning Us Military And Diplomatic Efforts In The Indo-Pacific Command Region*".

In a timely and insightful article Eugene Vertlieb, as translated by Dennis Faleris, writes from exile exploring Russian strategic conservatism and ideology of the former Soviet nation. In his work "*The Ideology of "Strategic Conservatism" from Russia's Imperial Perspective*" they highlight motivational factors through historical and cultural content which fuel modern Russian nationalism and patriotic ideology behind the 'special military operation' in Ukraine.

Shifting gears to the cyber domain many in the intelligence and security have heard the term 'blockchain' in reference to digital currency, but few have considered its potential applications in the defense community. As the U.S. works feverishly on technology forward future warfare capabilities including JADC2 and ABMS Dorian Belz presents a look at how blockchain technology can serve as a force multiplier in "*Blockchain: A Cyber Defense Force Multiplier*".

Our final article dips into the delicate world of HUMINT spy craft and the use of media outlets to frame strategic messaging. Ardavan Khoshnood presents the curious case of Lars Findsen, intelligence leaks, and prosecution in "*Intelligence Communities and the Media – The Case of the Danish Spymaster Lars Findsen*".

Global Security and Intelligence Studies strives to be the source for research on global security and intelligence matters. As the global threat-scape evolves over time, *GSIS* is evolving to keep pace. The Journal is enhancing its academic edge, impact, and reach. We are working to build stronger bridges between senior leaders, academics, and practitioners. In addition to new content that advances the global discussion of security and intelligence, readers can anticipate more special issues with a focus on current security concerns.

Global Security and Intelligence Studies is one of five journals sponsored by American Public University and published by Policy Studies Organization. The International Journal of Online Educational Resources (IJOER) publishes academic research with an emphasis on representing Open Educational Resources in teaching, learning, scholarship and policy. The Journal of Online Learning Research and Practice (JOLRAP) publishes articles that focus on aspects related to virtual instruction, technology integration, data, ethics, privacy, leadership, and more. Space Education and Strategic Applications (SESA) Journal encourages the publication of advances in space research, education and applications. And lastly The Saber and Scroll is a student and alumni led journal that publishes a variety of research on history or military history topics, book reviews and exhibit/museum reviews. Please visit https://www.apus.edu/academic-community/journals/index for more information on each journal.

A very special thank you to American Public University and the Policy Studies Organization for your generous and continued support of GSIS.

If you have research, notes, concepts, or ideas that you want to share please do not hesitate to reach out with your submission! Our editorial team is always available to support new authors seeking to make an impact on the industry. Please visit us on Scholastica, https://gsis.scholasticahq.com/for-authors, for specifics on submissions.

Carter Matherly, Ph.D.
Co-Editor in Chief

Matthew Loux, D.M.
Co-Editor in Chief

Jim Burch, D.M.
Associate Editor

Global Security and Intelligence Studies

Global Security and Intelligence Studies (GSIS) is published by the Policy Studies Organization on behalf of American Public University System.

Aims and Scope. GSIS is a bi-annual, peer-reviewed, open access publication designed to provide a forum for the academic community and the community of practitioners to engage in dialogue about contemporary global security and intelligence issues. The journal welcomes contributions on a broad range of intelligence and security issues, and from across the methodological and theoretical spectrum.

The journal especially encourages submissions that recognize the multidisciplinary nature of intelligence and security studies and that draw on insights from a variety of fields to advance our understanding of important current intelligence and security issues. In keeping with the desire to help bridge the gap between academics and practitioners, the journal also invites articles about current intelligence- and security-related matters from a practitioner perspective. In particular, *GSIS* is interested in publishing informed perspectives on current intelligence- and security-related matters.

GSIS welcomes the submission of original empirical research, research notes, and book reviews. Papers and research notes that explicitly demonstrate how a multidisciplinary approach enhances our theoretical and practical understanding of intelligence and security matters are especially welcome. Please visit : https://gsis.scholasticahq.com/for-authors.

Mikhail Gorbachev: A Transformational Leader

Andreea Mosila

Abstract

The peaceful end of the Cold War and the dissolution of the Soviet Union under the leadership of Mikhail Gorbachev surprised theorists of international politics, who failed to predict that the five decades-long conflict would end abruptly and peacefully. The element of surprise was Gorbachev, an uncharacteristic Soviet leader with Western thinking. His new thinking liberalized the Soviet system, brought historic changes to Eastern Europe, ended the Cold War, and dissolved the Soviet Union. To explain this chain of events, realists pointed to Gorbachev's foreign policy ideas causing the economic decline of the Soviet Union. Constructivists assumed that Gorbachev's plan to restructure the Soviet economic and political system and create a peaceful international order was the leading cause, while liberals blamed the liberalization of the domestic Soviet system and democratic peace. The peaceful end of the Cold War was heavily influenced by human behavior, a complex variable to predict. Gorbachev's upbringing in the province and his Moscow education created an unusual Soviet operational code: a leader interested in cooperation and peacemaking, development rather than destruction, and willing to pursue an interdependent foreign policy. This paper examines Gorbachev's psychological traits to explain his role in the peaceful end of the Cold War.

Keywords: Gorbachev, Cold War, Soviet Union, operational code, political psychology

Mikhail Gorbachev: Un líder transformativo

Resumen

El final pacífico de la Guerra Fría y la disolución de la Unión Soviética bajo el liderazgo de Mikhail Gorbachev sorprendieron a los teóricos de la política internacional, quienes no pudieron predecir que el conflicto de cinco décadas terminaría abrupta y pacíficamente. El elemento sorpresa fue Gorbachov, un líder soviético atípico con pensamiento occidental. Su nuevo pensamiento liberalizó el sistema soviético,

doi: 10.18278/gsis.7.1.2

trajo cambios históricos a Europa del Este, puso fin a la Guerra Fría y disolvió la Unión Soviética. Para explicar esta cadena de eventos, los realistas señalaron que las ideas de política exterior de Gorbachov causaron el declive económico de la Unión Soviética. Los constructivistas asumieron que el plan de Gorbachov para reestructurar el sistema económico y político soviético y crear un orden internacional pacífico era la causa principal, mientras que los liberales culparon a la liberalización del sistema soviético interno y a la paz democrática. El final pacífico de la Guerra Fría estuvo fuertemente influenciado por el comportamiento humano, una variable compleja de predecir. La crianza de Gorbachov en la provincia y su educación en Moscú crearon un código operativo soviético inusual: un líder interesado en la cooperación y el establecimiento de la paz, el desarrollo en lugar de la destrucción, y dispuesto a seguir una política exterior interdependiente. Este artículo examina los rasgos psicológicos de Gorbachov para explicar su papel en el final pacífico de la Guerra Fría.

Palabras clave: Gorbachov, Guerra Fría, Unión Soviética, código operativo, psicología política

米哈伊尔·戈尔巴乔夫：变革型领导者

摘要

冷战的和平结束和苏联在米哈伊尔·戈尔巴乔夫领导下的解体令国际政治理论家感到震惊，他们没有料到长达五十年的冲突会突然和平结束。震惊的要素在于戈尔巴乔夫，一位不寻常的、具有西方思想的苏联领导人。他的新思想解放了苏联体制，给东欧带来了历史性的变化，结束了冷战，解散了苏联。为了解释这一系列事件，现实主义者指向戈尔巴乔夫的外交政策理念，后者导致了苏联的经济衰退。建构主义者认为，戈尔巴乔夫在重组苏联经济体系和政治体系并建立和平国际秩序方面的计划是主要原因，而自由主义者则指责苏联国内制度的自由化和民主和平。冷战的和平结束在很大程度上受到人类行为这一难以预测的复杂变量的影响。戈尔巴乔夫在省内的成长经历和他在莫斯科接受的教育创造了一个不同寻常的苏联行动准则：一位对合作、和平、发展（而不是破坏）感兴趣并愿意奉行相互依存的外交政策的领导人。本文分析了戈尔巴乔夫的心理特征，以解释他在冷战和平结束中的作用。

关键词：戈尔巴乔夫，冷战，苏联，行动准则，政治心理学

I. Introduction

Between 1989 and 1991, the world witnessed some of its most significant historical moments. After five decades of nuclear proliferation and ongoing tensions between the two superpowers, the United States and the Soviet Union, the Cold War came to an unexpected and abrupt peaceful end, followed by the dissolution of the Soviet Union (USSR) under the rule of the Soviet president Mikhail Gorbachev, the key actor in the peaceful conclusion of this conflict. A transformational leader of the likes never seen before in the Soviet Union, Gorbachev's new way of thinking liberalized his country, brought historic changes to Eastern Europe, and ended the Cold War. That Gorbachev's ideas surprised the Western world and the Soviet Union is simply an understatement. The main theories of international politics failed to foresee the Cold War's peaceful end and the collapse of the USSR. The reasons this conflict—spanning almost five decades—ended peacefully and in less than two years became a debate among the theorists. From a realist perspective, Gorbachev's foreign policy was at the root of the economic decline of the Soviet Union. On the other hand, constructivists claimed that restructuring the economic and political Soviet system to create a peaceful international order, Gorbachev's "new thinking," or "Perestroika," was the cause that ended the Cold War (Snyder, 2005, 57). Liberals believed the Cold War unexpectedly ended because of Gorbachev's democratic peace proposal and his liberalization of the domestic Soviet system (Lebow, Risse-Kappen, 2001, 66). In reality, the end of the Cold War was heavily influenced by human behavior. Human behavior is difficult to predict. This theory application paper discusses Gorbachev's role in ending the Cold War and the collapse of the USSR. The paper focuses on Gorbachev's personality features that influenced his decision-making process and led to these significant events. The analysis is conducted from the perspective of realist, liberal, and constructivist theories. Image theory is also discussed.

II. Method

The method used in this theory application paper is to compare and contrast the research and theoretical views on the end of the Cold War to evaluate the consequential role of Mikhail Gorbachev's leadership. Personality traits, including motivation, belief, style, self-confidence, and cognition, are utilized to assess Gorbachev's personality at a distance to understand potential policy implications at the global level.

III. Analysis

1. Mikhail Gorbachev

Mikhail Gorbachev was born to a low-income family of peasants, in the small village of Privolnoe, close to the city of Stavropol in the southwest of the Soviet Union. As a child, he often had to help the family by working in the fields and starved for days on end. His childhood also coincides with some of the most challenging years of the USSR, marked by famine, collectivization, the aggressive politics of Joseph Stalin culminating with the Great Purge, and World War II. Gorbachev witnessed firsthand the injustice and impact of land collectivization, a fundamental cause of the 1932-1933 famine (Taubman, 2018, 7). His family's farm was collectivized, and several family members died of starvation. Stalin's Great Purge, the exclusion from the Communist Party of those unfaithful to the Stalinist doctrine, affected Gorbachev directly when his paternal grandfather was arrested in 1934 (Taubman, 2018, 18). World War II followed, and the German occupation of Privolnoe in 1941. His father had to join the Red Army on the frontlines of World War II, leaving the ten-year-old Mikhail in charge of his family and work on the farm, returning severely wounded in 1944 (Taubman, 2018, 26).

This traumatic period had a significant impact on Gorbachev's development later in life, his leadership style, and his desire to reform the domestic Soviet system to help everyday citizens. The direct effects of collectivization and famine on his family convinced Gorbachev that Stalin's agricultural collectivization was an injustice. As the Stavropol Communist Party secretary, Gorbachev reorganized the collective farms, improving the life of field workers by giving them more freedom to plan their work and increasing the size of individual land lots. Later, as the Secretary of the Central Committee, Gorbachev's agrarian reform offered collective farms more freedom to sell part of their production for a profit rather than surrendering the entire production to the Soviet state (Taubman, 2018, 53). Gorbachev could never overcome the trauma of his direct experience with World War II. Consequently, as the leader of the Soviet Union, Gorbachev showed unprecedented reluctance to use any military force. He oversaw the USSR withdrawal from Afghanistan and masterminded the withdrawal from the military and political intervention in Eastern Europe. As the leader of the Soviet Union, Gorbachev's changes to the Soviet system were so drastic and unexpected that he was later assessed as the first Soviet leader that acted as an "authentic Western politician" (Pop, 2011, 12). Political psychologists characterized Gorbachev as a "transformational leader" (Janis, 1989, 2).

Looking back on his early life, it is evident that Gorbachev's main psychological traits that played a significant role in his decision-making process originated in his life experiences. His early life in the province, combined with his later education in Moscow, led to a unique understanding of how central control affected the Soviet population (Janis, 1989, 1) and his desire to make the system better for everyone. The unique experience and traits of the Soviet leader were the main element of surprise and an important reason the theories in international politics failed to predict the peaceful end of the Cold War. As a post-world war leader, Gorbachev was more liberal in thinking and more willing to make significant decisions and changes to the system.

2. Realism, Liberalism, and Constructivism

The reason theories of international relations failed to predict the Cold War would end peacefully remains debatable. To make predictions, theorists look at the past and analyze behavior patterns. To evaluate the validity of any of these theories, it is essential to look at their success in predicting significant moments in history (Gaddis, 1992, 10). An integral theory approach is behavioralism, based on direct observation, employing the scientific method to col lect data. However, this method is relatively slow, and therefore behavioralists were not done collecting the necessary data by the time the Cold War ended. Their conclusions drawn from available data were simply tentative (Gaddis, 1992, 21). This may be one important reason why theories of international politics failed to predict how the Cold War would end. Gaddis (1992) blamed social sciences for trying to gain legitimacy as real science by using methods such as behavioralism based on scientific research methodologies. These failed to predict human behavior because human behavior is complex, and a model to simulate it would be challenging and impossible (Gaddis, 1992, 55). Kratochwil (1993) agrees that attempting to apply the scientific method to a theory of international relations is the wrong approach. The neo-realist and realist attempt to gain credibility by using the scientific method to obtain precise results in their assessment has failed and led to the inability to discover significant insights into the political situation of the Cold War (Kratochwil, 1993, 64).

Because the peaceful end of the Cold War surprised theorists of international relations, a custom-designed theory of Soviet foreign policy and its role in the domestic policy of the USSR may be needed to explain and understand the causes of this significant historical event of the twentieth century (Snyder, 2005, 55). Snyder (2005) used the unique method of connecting the assessments of the different theories, specifically the constructivist, realist, and neo-realist perspectives on what was at the root of ending the Cold War abruptly.

While constructivists believed that the end of the Cold war was attributed to ideas—in this case, Gorbachev's "new thinking"—realists claimed the end of the Cold War and the decline of the Soviet Union were caused instead by material factors—the economic fall of the USSR, a situation that influenced Gorbachev's foreign policy. This foreign policy was designed to save a declining Soviet Union (Snyder, 2005, 55). The element of surprise was Gorbachev himself, a counterrevolutionary leader who proposed an unexpected foreign policy that reformed the Soviet economic and political system. It is fair to say that the domestic goals of the Soviet leader are at the root of his foreign policy, which in turn put an end to the Cold War (Snyder, 2005, 55). Realists could not explain why Gorbachev's revolution went as far as it did to lead to the disintegration of the Soviet Union. In contrast, constructivists did not recognize that the reason behind the new foreign policy was not idealistic for practical purposes. Gorbachev's transformative foreign policy was designed to force domestic change in the Soviet system (Snyder, 2005, 56).

However, the realist and constructivist failures to assess the end of the Cold War are understandable. These are theories of international politics, and while the primary cause may have been a transformative foreign policy, the theories are not equipped to analyze domestic causes. Therefore, to connect the divide between realism and constructivism, a specific approach is necessary to explain how Gorbachev's foreign policy relates to the Soviet state and its domestic troubles (Snyder, 2005, 56).

A. Realism and Neo-Realism

When the Soviet Union collapsed, realists predicted a shift in global power from bipolarity to multipolarity, which meant more conflict and less stability. Between 1989 and 1991, realists expressed concern for the collapse of the bipolar world. Realism is based on the idea that global governance is anarchic. Therefore, stability on the global stage requires bipolarity, which explains the long peace after the Second World War through the existence of bipolarity in the international system: two superpowers balancing each other (Lebow, 1994, 252). To preserve superpower status, states tend to maintain their sphere of control and influence. This explains the failure of realism to predict the peaceful end of the Cold War and the dissolution of the USSR. Gorbachev's decision to give up control of the Eastern European block and withdraw the Soviet military from Afghanistan surprised realist theorists. The idea that a superpower would abandon its sphere of influence in Eastern Europe and its show of military power in Afghanistan cannot be reconciled with realism (Lebow, 1994, 262). Gorbachev's foreign policy was inconsistent with realist predictions, going as far as to accommodate its adversary, the United States.

Some realists claimed that Gorbachev's foreign policy was not, in fact, uncommon, but rather in accord with the realist idea that maintaining power is the priority. Gorbachev's domestic reform was meant to increase the superpower status of the USSR, and his foreign policy was designed to simply support his domestic agenda (Lebow, 1994, 263). However, this assumption is inconsistent with the fundamental realist idea that superpowers would not renounce their sphere of influence. How would a foreign policy supporting a domestic agenda lead to a decline of a superpower? (Lebow, 1994, 263). While Gorbachev spoke extensively about significant economic reform, he delayed any measures to do anything about the economy, taking very few steps in this direction. At the same time, his foreign policy plans were implemented fast and steady. To claim his foreign policy was designed to support the domestic policy is not a persuasive explanation (Lebow, 1994, 266).

The approach of neo-realism to the end of the Cold War was described as "embarrassed" because the expectations and predictions of this theory were not factoring in peace, but hegemonic war (Kratochwil, 1993, 63). The fact that the domestic reform generated changes at the international level to the Soviet system was also a surprise. Neo-realism is a theory based on power and unable to comprehend power renunciation. The attempt of neo-realism and realism to gain credibility by using the scientific method to get precise results in their assessment has failed and led to the inability to discover significant insights into a political situation that realism could offer (Kratochwil, 1993, 64).

Approaching international politics from a scientific perspective led to a lack of significant questions and the abandonment of traditional methods of assessing human behavior. Since neo-realism has failed to predict the changes of Perestroika, an alternative method of analysis is needed in international politics. Neo-realism predictions failed because Gorbachev's actions, decision-making style, and human behavior cannot be quantified. Instead, by focusing on leaders' political action and behavior, this new framework of analysis would be able to better predict future changes on the global stage (Kratochwil, 1993, 80). Koslowski (1994) agrees with the assessment that Gorbachev's actions undermined neo-realist predictions of how the Cold War would end. Gorbachev was able to reverse the Bolshevik conquest of domestic Soviet politics and the Soviet dominance in Eastern Europe by simply ending the Brezhnev Doctrine (Koslowski, 1994, 216). This surprised neo-realism because this theory expected bipolarity to last, and Gorbachev's actions led to the disintegration of the Soviet Union, and hence the end of bipolarity in world power. Besides this, the changes that led to the end of the Cold War did not match any of the neo-realist predictions, as these changes did not occur as a consequence of a hegemonic war, nor because a new superpower would arise or gaps in military capabilities (Koslowski, 1994, 217). On the contrary. Gorbachev's action to discard the Brezhnev Doctrine led to the overthrow of the communist regimes in the Eastern European countries and the end of the Warsaw Pact. Neo-realists' failure is rooted in their disregard for the national politics of the Soviet Union and claim that the peaceful end of the Cold War was a result of Gorbachev's desire to reform the domestic Soviet system. Unlike his predecessors, Gorbachev understood that political and economic reform within the USSR could only occur in a global environment of peace. For this purpose, maintaining dominance in Eastern Europe through the use of force and participating in an ongoing Cold War with the West was counterproductive (Koslowski, 1994, 218).

Another failure of neo-realism was its prediction that the opponent, the United States, would take advantage of its adversary's weaknesses, a Soviet economy and political system in decline. It would implement an aggressive foreign policy that would further weaken the USSR. However, the United States chose cooperation, invited the USSR to join international organizations, and went even further to offer financial aid to help with their economic reform. Neo-realism's explanation for this behavior was that a superpower would do anything to avoid a power vacuum on the global stage, which is the reason behind the U.S.'s help. However, this argument is inaccurate because it contradicts the neo-realist idea of power maximization (Koslowski, 1994, 220). Furthermore, the United States continued multilateral cooperation even after the USSR collapsed.

B. Liberalism

While realism asserts that the international system is anarchic and prone to conflict, liberalism is based on cooperation rather than conflict, freedom, and democratic peace, with a psychological root in the retreat of fear (Hymans, 2010, 464). An ongoing cold war with the West was a significant obstacle to Gorbachev's idea to liberalize and decentralize the Soviet system. Long-term peace and the global integration of the Soviet Union was much more critical to achieving a stable, centralized economy. Gorbachev's decision to end any military operations in Afghanistan and withdraw from a military and political intervention in Eastern Europe were actions rooted in liberal principles (Snyder, 2005, 59). As a post-world war leader, Gorbachev was more liberal in thinking than the Soviet leaders that came before him, and he was not afraid of the consequences of significant decisions.

C. Constructivism

Constructivism considers the international system an artificial collection of institutions—a collection that includes states and other actors on the global stage. Changes to this system usually occur within established and accepted conventions within this artifice. However, if these conventions are altered, more fundamental system changes can occur (Lebow, 2001, 99). This means that one leader can, in theory, change the entire system. Constructivists claimed that Gorbachev's reform of the Soviet economic and political system with the scope to create a peaceful international order, the "new thinking," "Perestroika," was the leading cause of the peaceful end of the Cold War. Ideas and agency are at the root of this unexpected event in history (Snyder, 2005, 57). However, constructivism could not recognize that Gorbachev's "new thinking" was to use foreign policy to make systemic changes within the Soviet Union (Snyder, 2005, 56). Constructivism explains Gorbachev's abandonment of the Eastern European block not as a decision to give up control but rather to retain it through the domestic reform of the communist system, hoping that the Perestroika model would prevail and inspire other communist leaders to follow.

While psychological realism claims that fear is the foundation of a conflict in the international system, psychological constructivism sees spirit and honor at the root of an armed conflict that aims to establish a hierarchy (Hymans, 2010, 463). However, it is understandable why this honor-based theory may have failed to predict the peaceful end of the Cold War. The main limitation of an honor-driven system is that, in a world where honor plays a significant role, order in the global system is often established through armed conflict to create a hierarchy on the international stage and not for protection (Hymans, 2010, 463). Gorbachev's choice to renounce the Soviet sphere of influence in Eastern Europe and pursue military withdrawal from Afghanistan did not match the idea that a conflict is at the root of any effort to establish hierarchy on the global stage. The constructivist assessment that at the root of ending the Cold War was Gorbachev's new thinking was accurate. However, it does contradict its idea of the spirit and honor-based world.

3. *Image Theory*

Image theory explains well why the peaceful end of the Cold War was so unexpected. Belief systems affect perception and decision-making in international politics. Leaders develop an image of their adversaries based on a belief system. The decision-making process in international politics is affected by this image that often is not a real, objective one but rather an image based on stereotypes (Holsti, 1962, 3). National stereotypes are concerning in international politics, as they can influence conflict and the perception of an actor as being "bad" or the "enemy." In a global conflict, decision-making actors tend to act according to the image of the adversary they have established based on their perception. Therefore, conflicts often occur between distorted images of states and not reality (Holsti, 1962, 3). Applying these ideas to the Cold War, the West developed a belief system based on a stereotypical image of the Soviet leader embodied by Stalin, Kruschev, and Brezhnev. These were leaders committed to the strength of the socialist system both within the Soviet Union and at the foreign policy level through actions such as military interventions in countries members of the Warsaw Pact to ensure the survival and thriving of the socialist governments. Gorbachev's leadership did not fit the pattern, the belief system, or the stereotypical image of the Soviet leader. The leadership of Mikhail Gorbachev came as a pleasant surprise to the West. Gorbachev's decision-making in reorienting the Soviet foreign policy towards reconciling with the West defied the accepted operational code of the Soviet leader. The unique traits of Gorbachev's personality were the element of surprise, as Gorbachev was a transformational leader (Janis, 1989, 2) rather than the classic Soviet leader predicted by image theorists.

Foreign policy decisions are based on the world's image and not the real world, which is quite the issue when such foreign policy decisions are based on a false image. Making decisions based on a manufactured image of the external world has been known as the "operational code" (George, 1969, 191). It is essential to understand the belief systems that lead to a specific operational code to analyze decision-making in international politics. This will help better understand leadership style (George, 1969, 221). Interestingly, despite the failure of international relations theories to predict the peaceful end of the Cold War, theorists concluded that it is essential to analyze leaders' subjective belief systems and behavior to better predict future decisions in foreign policy (Schafer & Walker, 2006, 3). Analysis of the operational code became an essential method in political psychology and international relations. The theory of the operational code was a product of the early Cold War and had as its scope the study of political leaders to predict the intentions and actions of Soviet officials (Gaddis, 1992, 9).

4. Gorbachev's Operational Code: A Behavior Analysis

Personality traits such as motivation, belief, style, self-confidence, and cognition can be utilized to assess a political leader's personality at a distance to understand potential policy implications at the global level. For this purpose, it is helpful to have an operational code focused on Gorbachev's view of self and the relations of the self with the world and its politics (Winter, 1991, 224). Gorbachev's interviews and speeches were compared with previous Soviet leaders, including Stalin, Khrushchev, Brezhnev, and American presidents George Bush and Richard Nixon. The analysis was based on interpersonal style, cognition, belief, and power motivation and pointed to an unexpected similarity between Gorbachev and Bush and Nixon and dissimilarity between Gorbachev and Stalin, Khrushchev, and Brezhnev. This comparison suggests potential issues or problems a leader might encounter (Winter, 1991, 230).

Advances in neuroscience reveal new details about how the brain functions and can help assess how emotion affects political decisions and can be used to predict future choices (McDermott, 2004, 691). Emotion does matter in determining personality, and therefore, it cannot be ignored. It is essential in quick decision-making; analyzing it can reveal necessary information about a leader and the situation in which a specific decision was made (McDermott, 2004, 702). Emotions and emotional relationships and the disregard for these can be helpful in the assessments of international relations theories. The disregard for emotion can cause the failure of peacebuilding following wars as, most of the time, these attempts can trigger "emotional fire" (Crawford, 2000, 116). This ignorance is based on the assumption that state actors are rational, even if many of these theories are based on emotion, for example, realism on fear (Crawford, 2000, 116). Emotions such as fear, anger, or empathy deserve some attention and study in international politics.

Personality psychologists used emotion, achievement, affiliation, and power interest to measure Gorbachev's motive. Beliefs such as nationalism and trust are used to measure authoritarian personality tendencies. Self-confidence is used to assess if a leader would be active or reactive (Winter, 1991, 222). Besides these personality traits, Gorbachev's operational code could denote if he perceived the political environment as hostile or friendly, what his historical legacy would be, how he could achieve his political goals, and what means he should use to achieve these aspirations (Winter, 1991, 223). The analysis reveals a motive profile characterized by high achievement, high affiliation, average power interest, and a rationally cooperative personality. The high affiliation score denotes a leader that would pursue agreements on arms limitations. Gorbachev scored lower in power motivation than other Soviet leaders. However, compared to George Bush and Richard Nixon, the power motivation result was similar and average. Lower to average power motivation means the leader will be less likely to use force and aggression to accomplish his goals (Winter, 1991, 231).

The beliefs and styles analysis showed a similarity between Bush and Gorbachev again, as both scored high on nationalism and distrust of others (Winter, 1991, 231). While usually high nationalism means very simplistic thinking, in Gorbachev's case, a heightened nationalism and distrust score and an increased complexity score suggest a refined ability to differentiate and understand different points of view, principles, and policies and use the discrepancies to develop complex generalizations. High complexity can mitigate heightened nationalism and distrust through intellectualization. Gorbachev's high confidence score means an optimistic personality (Winter, 1991, 231). As far as traits, Gorbachev showed a dynamic interpersonal style, was very expressive but in control of emotions, displayed low anxiety, was sensitive to criticism but knew how to control the challenge, and was overall an excellent actor-politician (Winter, 1991, 235). As far as decision-making is concerned, Gorbachev's personality showed a relatively moderate than impulsive tendency, ability to utilize the ideas and solutions of others to solve a problem, and a stable extrovert (Winter, 1991, 235).

Adding up these findings, Gorbachev's operational code is determined by his characterization as a friendly, optimistic leader with broad goals and vision, positive reactions, focused on words rather than actions, and virtue rather than threats. This mix of personality traits makes a leader capable of exerting great control of foreign policy and foreign policy outcomes (Winter, 1991, 234). Gorbachev's motivation was based on achievement and affiliation, not power and exploitation, which explains his tendency to seek the help and ideas of others in achieving his goal of solving national problems and bettering the Soviet system. This operational code describes a leader prone to cooperation, peacemaking, and interest in development rather than destruction, with a tendency to pursue an interdependent foreign policy.

As a leader who made a significant difference in history because of his unique personality capable of changing the international system and global order, Gorbachev can be compared to Napoleon and Churchill, primarily through his sound decision-making under stress while resisting social pressure, together with his lack of strict adherence to the Bolshevik operational code was critical to his success (Janis, 1989, 3). His renowned popularity abroad compares to that of the Egyptian president Anwar Sadat, who was extensively admired outside of Egypt, but ultimately fell prey to the divisiveness of his own country, which eventually led to Sadat's inability to achieve his visions for Egypt (Post, 1989, 2).

Public statements Gorbachev made between 1985 and 1991 were used to analyze his behavior patterns and how the external environment affected these patterns and reveal the relationship between the crises he faced, the policies he proposed to solve these crises, and his decision-making style (Wallace et al., 1996, 454). Gorbachev exhibited a different behavior pattern from his first years in power than the previous Soviet leaders. Still, this behavior pattern developed over time to grow in complexity when dealing with foreign policy issues while remaining simplistic when dealing with domestic problems. This could be why Gorbachev was a successful leader on the global stage but was seen as a failure domestically by the Soviets (Wallace et al., 1996, 454).

External environmental factors play an essential role in shaping a leader's behavior and complexity. For example, right at the beginning of his tenure in power, in 1985, Gorbachev faced severe opposition from conservative Soviet leadership when he proposed simultaneously a critical change in foreign policy, the reduction of strategic warheads by half, and radical economic reform in domestic policy. As a reaction to this opposition, Gorbachev removed some of the Soviet leaders holding positions in the council of ministers and the military command for too long (Wallace et al., 1996, 458). This measure was sufficient to signal to the West that Gorbachev was a different kind of Soviet leader. Gaining a reputation as a transformational leader, Gorbachev continued to score successes on the international stage. However, that cannot be said of the domestic stage. While successfully establishing and strengthening his power within the communist party, an economy in free fall, budget deficits and nationalist movements within the Soviet Union fighting for the independence of Soviet states were not good news for Gorbachev (Wallace et al., 1996, 459). And yet his desire for domestic change remained unwavering.

Wallace's study found significant differences in domestic and foreign policy behavioral complexities. Interestingly, Gorbachev's foreign policy strategies were complex and versatile in dealing with the changes on the global stage at that time. In contrast, his domestic policies remained simplistic and failed to implement the domestic reform of his desire and even help him stay in power (Wallace et al., 1996, 468). While Gorbachev was a unique Soviet leader in foreign policy, he was not different domestically from other Soviet leaders before him, unable to keep the Soviet economic decline under control.

5. Gorbachev's Decision-Making Analysis

To adequately explain the significant changes in the international systems that led to the end of the Cold War, it is crucial to give credit to the world leaders involved in the event, particularly

Gorbachev, and how their decisions were influenced by the events of the day. An excellent example of this is the impact of the Prague Spring, an event that took place in 1968, long before Gorbachev came into power, on the future decisions of this Soviet leader (Suri, 2002, 77). The Prague Spring was the invasion of Czechoslovakia by the Soviet army, another military action that affected young Gorbachev. As already discussed, Gorbachev's life was significantly influenced by World War II, and because of his personal experience with war, he became a strong opponent of any form of violence. He writes in his memoirs about the influence of the Prague Spring on his future decisions as a leader and how events like this made him into a reformist thinker (Suri, 2002, 77). His experience with war made him realize that to accomplish his domestic economic reforms, he would need to be able to operate in the context of a peaceful international environment. An ongoing Cold War would have meant economic stagnation at best and complete economic decline at worst, precisely the domestic situation Gorbachev wanted to change (Suri, 2002, 78). Successful cooperation between the East and the West was essential for the USSR to allocate all its resources to solving domestic issues rather than maintaining the needs for an ongoing Cold War (Suri, 2002, 78).

Gorbachev's Perestroika was also rooted in a new philosophical approach for a Soviet leader. While previous Kremlin leaders focused on a class approach, Gorbachev favored "humanistic universalism" (Suri, 2002, 79). As an avid reader and scholar of politics and philosophy, this is not surprising. Emphasizing common ground between the East and the West regarding values rather than differences, Gorbachev built his foreign policy on this humanistic universalism approach, proposing that common ground can overcome the differences at the root of the Cold War (Suri, 2002, 79). While this philosophical approach was welcomed on the global stage, within the USSR, some communist party leaders did not favor a humanistic direction for the Soviet system but rather the need for a radical change. But as Secretary of the Central Committee, Gorbachev chose to surround himself with "new thinkers" willing to favor cooperation rather than conflict (Suri, 2002, 80). The domestic "new thinking" was crucial for developing the foreign "new thinking," and without it, an improvement of the relationship between the East and the West was inconceivable. Domestic changes and development were vital to ending the tensions at the global level and replacing them with cooperation and long-term peace (Suri, 2002, 80). Another critical factor in ending the Cold War was Gorbachev's decision to give up control of the Eastern European block. It remains unknown how Gorbachev made this decision and how he could convince his fellow communist party leaders to agree and renounce a large sphere of influence for the USSR (Suri, 2002, 82). However, withdrawal from political control of the Eastern Europe countries, just like the military withdrawal from Afghanistan, are behavior patterns exhibited by Gorbachev because of his experience with war, invasions, and foreign control. How could he ask for international cooperation while exerting control over several independent nations?

Gorbachev's attempt to reform the socialist system was the second such attempt, following that of Nikita Khrushchev, with the significant difference that Gorbachev's action was a fatal hit to the Soviet empire in particular and the socialist system in general and precipitated the end of the Cold War (Pop, 2011, 12). The problem with Gorbachev's attempt to reform the Soviet system was that it took place when the system was already in free fall, and a fundamental reform at such a moment was dangerous. In the words of Alexis de Tocqueville, "the most perilous moment for a bad government is one when it seeks to mend its ways" (Pop, 2011, 12). Consequently, the attempt to fundamentally reform the system led to its disintegration and the collapse of the USSR. Gorbachev's critics and admirers alike attributed this to the so-called "Gorbachev factor," a set of personal traits including optimism, naivete, self-confidence, and the desire to act ad hoc when needed, while also postponing difficult decisions, his Western thinking, but also the aversion towards the use of force (Pop, 2011, 13).

For example, Gorbachev's tendencies to delay difficult decisions were at the root of his failure to implement economic reform in time, while his naivete was at the heart of his belief that Eastern Europe could maintain socialism as a form of government. Consequently, he was not worried when the regimes of the Eastern European block began to disintegrate. However, his Western thinking was one of his most admired traits on the global stage, and it led to his assessment as the first Soviet leader that acted as an "authentic Western politician" (Pop, 2011, 12). The criticism within the Soviet Union and the admiration abroad determined Gorbachev to turn more and more toward the West.

To evaluate the role the "Gorbachev factor" played in ending the Cold War and the collapse of the communist regimes in Eastern Europe, including the dissolution of the USSR, it is vital to question why these processes did not occur before 1989 (Pop, 2011, 13). Before the mid-80s, the idea of withdrawing politically from the Eastern European block was not vehiculated at all in the Soviet Union. On the contrary, there was a precise determination to maintain this sphere of influence in the area even if force was needed. A trend to abandon socialism existed in the Eastern European countries way before 1989. Still, the USSR continued to firmly maintain its hegemony in the area under the so-called "Brezhnev Doctrine," the idea that the USSR had the right and obligation to intervene in any Eastern European country to preserve the socialist system (Pop, 2011, 14). Gorbachev decided to renounce the Brezhnev Doctrine, opening the door to geopolitical and ideological change in Eastern Europe. Gorbachev improved relations with the West, making the abandonment of the Stalin–Churchill "Percentage Agreement" of 1944 possible. Before Gorbachev, this agreement kept the West from interference in Eastern European affairs, a significant obstacle in eliminating the socialist system and implementing capitalism and democracy in these countries (Pop, 2011, 14). With Gorbachev in power, in the second half of the 1980s, a new relationship developed between the East and the West, allowing for European integration of the Eastern European countries. Under Gorbachev's leadership, the final meeting of the Warsaw Treaty Organization in December 1989 agreed that the 1968 military intervention in the Prague Spring was illegal (Pop, 2011, 15). Reform in Eastern Europe was only possible once the policy of non-intervention was agreed upon by all members of the Warsaw Treaty Organization.

Gorbachev's Perestroika aimed to restructure and revitalize the Soviet Union, render its economy competitive on the global stage, and improve the image of socialism in the world (Patman, 1999, 578). Gorbachev's leadership was undoubtedly at the root of this new way of thinking, together with the ongoing decline of the Marxist-Leninist political system and an adversary (the U.S.) with an ongoing program of military renewal that forced Soviet leadership to choose between an ongoing cold war with a far superior military adversary, or cooperation (Patman, 1999, 579). As a post-Stalin era Soviet leader, Gorbachev was able to introduce a new foreign policy and arms control while minimizing the power of the domestic Stalinist leaders, reforming Stalinist institutions, and implementing an organic involvement of the Soviet Union in the global economy (Snyder, 1987, 95).

IV. Conclusion

Mikhail Gorbachev was a key actor in the peaceful conclusion of the Cold War. A one-of-a-kind transformational leader, Gorbachev's new way of thinking liberalized the Soviet Union, brought historic changes to Eastern Europe, and ended the Cold War. But Gorbachev's ideas surprised both the Western world and the Soviet Union alike, leading to the main theories of international politics failing to foresee the Cold War's peaceful end and the collapse of the USSR. Realists claimed Gorbachev's foreign policy was at the root of the economic decline of the Soviet Union. Liberals believed the Cold War unexpectedly ended because of Gorbachev's democratic peace proposal and his liberalization of the domestic Soviet system. Constructivists suggested that restructuring the economic and political Soviet system to create a peaceful international order, Gorbachev's "new thinking," or "Perestroika," was the cause that ended the Cold War.

Ultimately, the peaceful end of the Cold War was heavily influenced by human behavior. This theory application paper analyzed Gorbachev's role in ending the Cold War and the collapse of the USSR by looking at Gorbachev's personality features that influenced his decision-making process and led to these significant events. The analysis was conducted from the perspective of realist, liberal, constructivist, and image theories. Personality traits, including motivation, belief, style, self-confidence, and cognition, were utilized to assess Gorbachev's personality at a distance to understand potential policy implications at the global level. Gorbachev's operational code, defined by his characterization as a friendly, optimistic leader with broad goals and vision, positive reactions, focused on words rather than actions, and virtue rather than threats, revealed a leader capable of exerting great control of foreign policy and foreign policy outcomes. This operational code revealed a leader prone to cooperation, peacemaking, and interest in development rather than destruction, with a tendency to pursue an interdependent foreign policy. This was difficult to predict for any international relations theories. Gorbachev and his decision-making in reorienting the Soviet foreign policy toward reconciling with the West to change the domestic landscape of the Soviet Union were the elements of surprise.

Bibliography

Adrian Pop. "Factorul Gorbaciov." *Sfera politicii* 19, no. 11 (2011): 12–.

Crawford, Neta C. "The Passion of World Politics: Propositions on Emotion and Emotional Relationships." *International security* 24, no. 4 (2000): 116–156.

Gaddis, John Lewis. "International Relations Theory and the End of the Cold War." *International Security* 17, no. 3 (1992): 5–58.

George, Alexander L. "The 'Operational Code': A Neglected Approach to the Study of Political Leaders and Decision-Making." *International studies quarterly* 13, no. 2 (1969): 190–222.

Holsti, Ole R. "The Belief System and National Images: A Case Study." *The Journal of Conflict Resolution* 6, no. 3 (1962): 244–252.

Hymans, Jacques E.C. "The Arrival of Psychological Constructivism." *International Theory* 2, no. 3 (2010): 461–467.

Irving Janis. "Psyching Out Gorbachev; The Man Remains a Mystery–Even as He Reshapes the World; His 'Iron Nerves' Can Solve Problems: FINAL Edition." *The Washington Post*. Washington, D.C: WP Company LLC d/b/a The Washington Post, 1989.

Jerrold Post. "Psyching Out Gorbachev; The Man Remains a Mystery–Even as He Reshapes the World; He Embodies a New Generation of Russians: FINAL Edition." *The Washington Post*. Washington, D.C: WP Company LLC d/b/a The Washington Post, 1989.

Koslowski, Rey, and Friedrich V. Kratochwil. "Understanding Change in International Politics: The Soviet Empire's Demise and the International System." *International Organization* 48, no. 2 (1994): 215–47.

Kratochwil, Friedrich. "The Embarrassment of Changes: Neo-Realism as the Science of Realpolitik without Politics." *Review of International Studies* 19, no. 1 (1993): 63–80.

Lebow, Richard Ned, and Thomas Risse-Kappen. *International Relations Theory and the End of the Cold War*. New York: Columbia University Press, 2001.

Lebow, Richard Ned. "The Long Peace, the End of the Cold War, and the Failure of Realism." *International Organization* 48, no. 2 (1994): 249–77. http://www.jstor.org/stable/2706932.

McDermott, Rose. "The Feeling of Rationality: The Meaning of Neuroscientific Advances for Political Science." *Perspectives on Politics* 2, no. 4 (2004): 691–706.

Patman, Robert G. "Reagan, Gorbachev and the Emergence of 'New Political Thinking.'" *Review of International Studies* 25, no. 4 (1999): 577–601.

Schafer, M, and S. Walker. *Beliefs and Leadership in World Politics: Methods and Applications of Operational Code Analysis*. New York: Palgrave Macmillan US, 2006.

Snyder, Jack. "The Gorbachev Revolution: A Waning of Soviet Expansionism?" *International Security* 12, no. 3 (1987): 93–131.

Snyder, Robert S. "Bridging the Realist/Constructivist Divide: The Case of the Counterrevolution in Soviet Foreign Policy at the End of the Cold War." *Foreign Policy Analysis* 1, no. 1 (2005): 55–72.

Suri, Jeremi. "Explaining the End of the Cold War: A New Historical Consensus?" *Journal of Cold War Studies* 4, no. 4 (2002): 60–92.

Taubman, William. *Gorbachev: His Life and Times*. W. W. Norton & Company. 2018. Kindle edition.

Wallace, Michael D., Peter Suedfeld, and Kimberly A. Thachuk. "Failed Leader or Successful Peacemaker? Crisis, Behavior, and the Cognitive Processes of Mikhail

Sergeyevitch Gorbachev." *Political Psychology* 17, no. 3 (1996): 453–72.

Winter, David G., Margaret G. Hermann, Walter Weintraub, and Stephen G. Walker. "The Personalities of Bush and Gorbachev Measured at a Distance: Procedures, Portraits, and Policy." *Political Psychology* 12, no. 2 (1991): 215–45.

Andreea Mosila is currently pursuing a doctorate in Global Security and holds an MA in Political Science and an MS in Aeronautics. Her primary area of research includes global governance crisis management. Highlights from her research include analyzing global leaders' psychological rationales to understand better the response to a crisis and propose strategies for better global management in a future crisis. She welcomes opportunities for continued research and collaboration: andreeamosila@gmail.com.

Coming Together: Strengthening the Intelligence Community Through Cognitive Diversity

David Kritz

Abstract

In the 2019 National Intelligence Strategy of the United States of America, the word *integrate*, or a variation of the word, was depicted 51 times. Of the seven enterprise objectives set by the Director of National Intelligence, the first two focus on integration of mission and business management. The purpose of this article is to demonstrate that cognitive diversity can both enable and inhibit cooperation, is dependent upon leadership style, and can lead to real-world implications that move from strategy to execution. If the Intelligence Community has a cognitively diverse workforce, then innovation is more likely to occur and remain sustainable. A qualitative methodology was used to address the question: *How can cognitive diversity increase integration endeavors across the 18 agencies to achieve the Intelligence Community's Vision and support the Mission?* Findings substantiate the importance of cognitive diversity to problem solve, work through complexity, and improve decision-making during times of crisis.

Keywords: Cognitive Diversity, Complexity, Diversity, Intelligence, Integration, National Security

Unidos: Fortalecimiento de la comunidad de inteligencia a través de la diversidad cognitiva

Resumen

En la Estrategia Nacional de Inteligencia de los Estados Unidos de América de 2019, la palabra integrar, o una variación de la palabra, se representó 51 veces. De los siete objetivos empresariales fijados por el Director de Inteligencia Nacional, los dos primeros se centran en la integración de la misión y la gestión empresarial. El propósito de este artículo es demostrar que la diversidad cognitiva puede habilitar e inhibir la cooperación, depende del estilo de liderazgo y puede tener implicaciones en el mundo real que van desde la estrategia hasta la ejecución. Si la comunidad de inteligencia tiene una fuerza laboral cognitivamente diversa, entonces es más probable que ocurra

doi: 10.18278/gsis.7.1.3

la innovación y siga siendo sostenible. Se utilizó una metodología cualitativa para abordar la pregunta: ¿Cómo puede la diversidad cognitiva aumentar los esfuerzos de integración entre las 18 agencias para lograr la Visión de la Comunidad de Inteligencia y apoyar la Misión? Los hallazgos corroboran la importancia de la diversidad cognitiva para resolver problemas, superar la complejidad y mejorar la toma de decisiones en tiempos de crisis.

Palabras clave: Diversidad Cognitiva, Complejidad, Diversidad, Inteligencia, Integración, Seguridad Nacional

聚集在一起：通过认知多样性加强情报界

摘要

美国《2019国家情报战略》中，“整合”一词或该词的变体被描述了51次。在国家情报总监制定的七项事业目标中，前两项聚焦于使命与业务管理的整合。本文旨在证明，认知多样性既能促进也能抑制合作，其取决于领导风格，并且能导致从战略到执行的现实世界影响。如果情报界拥有认知多样化的劳动力，那么创新更有可能发生并保持可持续性。使用定性方法研究一个问题：认知多样性如何增加18个机构的整合举措，以实现情报界的愿景并维护使命？研究结果证实了认知多样性对于解决问题、克服复杂性以及在危机时期提升决策的重要性。

关键词：认知多样性，复杂性，多样性，情报，整合，国家安全

Can we talk of integration until there is integration of hearts and minds? Unless you have this, you only have a physical presence, and the walls between us are as high as the mountain range.

—Chief Dan George (Chief of the Tsleil-Waututh Nation)

Introduction

If one thing continues throughout the 21st Century, it will be change. It is difficult to pin down the specificity of change as it is literally and figuratively occurring all around us in many ways. An example is the repercussions for remaining stagnant during COVID-19, as businesses that did not evolve likely closed. Within the first year of the pandemic, approximately

200,000 U.S. businesses closed (Simon 2021). Change will emerge in predictable patterns, unpredictably by misinformation or disinformation, and perhaps a black swan or two will appear during an inconvenient time. Al-Asadi, Muhammed, and Dzenopoljac (2019) adroitly stated "since the postindustrial era, significant technological changes, innovation, and creativity have become the main pillars of sustainable competitive advantage of modern companies" (472).

If there is an emerging change to the way the world is perceived to work, then there needs to be an evolving and refinement with the skills employees possess if the institution is to remain relevant. This hypothesis statement is not only reflective of for-profit companies, but also includes government organizations, especially the Intelligence Community (IC). COVID-19 was a stark reminder for the IC that the value of intelligence does not lie with secrecy and working inside sensitive compartmented intelligence facilities, but finding truth to aid policymaker's decisions. Global pandemic withstanding, the IC cannot close and temporarily adjust to working outside traditional settings. It is also challenging to pin down the specifics of changes the IC faces as the threats and opportunities continue to evolve. This is depicted within the Office of the Director of National Intelligence (ODNI's) 2022 Annual Threat Assessment of the U.S. IC, which has eight crucial topics ranging from specific countries to Health Security, climate change, and conflicts and instability.

As the world continues to evolve in complexity, there is no organization held more accountable to changes across a wide spectrum of areas than the IC. This is due to the valid reason of acquiring the dual responsibilities of informing policymakers for national interest endeavors and advising through intelligence to defend the nation. With the recent addition of the Space Force, the IC is now comprised of 18 organizations and agencies, with a National Intelligence Program budget appropriated in 2020 for $62.7 billion. How the organizations within the IC interact to improve its mission is still occurring; however, due to the evolving landscape of threats and opportunities, the IC must evolve (Barger 2004). It is not lost on the senior leaders within the IC federation that change is needed to thwart adapting threats, maintain relevancy, and mitigate intelligence failures such as an attack against the nation. One harsh critique that sums up some of the previous perceived intelligence failures comes from a Pulitzer-winning author Richard Rhodes, who stated that "for all its accumulating trillions in national treasure, failed to predict 9/11, predicted Iraqi WMD that weren't there, and in 'the most glaring example ... the colossal intelligence failure of 2011,' failed to predict the Arab Spring" (2011, n.p.).

To begin, we should acknowledge that the drive for change and integration emerged from the terrorist attacks of September 11, 2001 (9/11). The IC was vastly different two decades ago in terms of organization and mindset. From the Cold War up to 9/11, the IC did not task for, nor have the awareness, to appreciate integration across unity of effort as each agency focused on specialized missions to collect upon and then disseminate in finished and often classified reports (The National Commission on Terrorist Attacks Upon the United States 2004). In the aftermath, the IC went through a substantial change with the establishment of the Office of the Director of National Intelligence and the Department of Homeland Security. Transformational change in diversity was required to gain more profound knowledge in understanding global ethnic and cultural differences (Miner and Temes 2022).

Barger (2004), a RAND Intelligence Policy Center's 2003 IC Fellow, argues from the perspective of insiders claiming that while reform causes increased bureaucracy and regulation there is a paucity of increased intelligence performance. Perhaps, but perhaps not. Lingering attitudes of older workforce generations may include perceptions of change as being punitive and opening the aperture of building relationships across peer organizations within the IC is suspect. More egregious and hopefully uncommon event as suggested by Best (2011) is the attempt to prevent other IC organizations from the perception of one organization gaining leverage over another to get ahead through the willful decision not to cooperate in the sharing of information. The IC suffered from a dated Cold War mentality that inhibited collaboration and as a result the terrorist attacks on 9/11 demonstrated the IC needed to evolve in multiple areas to include increased communication and integration (Hamrah 2013). This mindset continues as the risk of complacency and not adapting is likely to lead to grave consequences from groupthink as a slippery slope. If threats toward national interests and the areas to communicate opportunities for decisionmakers evolve, then the IC needs to be agile and adaptive. This endeavor occurs with the model presented herein developed by the author of multiplying communication, cooperation, and collaboration to yield integration.

How the IC will implement change is depicted by the strategic direction provided by the Director of National Intelligence (DNI) across this federation every four years in the form of the National Intelligence Strategy of the United States of America. Of the seven enterprise objectives set by the DNI, the first two focus on integration of mission and business management. One method on how the IC will meet the DNI's primary objective of Integrated Mission Management is to "strengthen and integrate IC governance bodies to increase transparency, prioritize and optimize resources, balance tradeoffs, and manage risk" (National Intelligence Strategy 2019, 18). The DNI's second enterprise objective is Integrated Business Management. Within this objective the IC's business functions and practices include "the coordinated development, alignment, de-confliction, execution, and monitoring of strategies, policies, plans, and procedures needed to manage and secure the IC and its people, information technology, and physical infrastructure" (National Intelligence Strategy 2019, 19). The DNI's objectives are aligned under the IC's vision and mission statement. The National Intelligence Strategy defines the IC Vision as "a Nation made more secure by a fully integrated agile, resilient, and innovative Intelligence Community that exemplifies America's values" (3) and the IC Mission to "provide timely, insightful, objective, and relevant intelligence and support to inform national security decisions and to protect our Nation and its interests" (3).

As the DNI charged the IC to fulfill customer requirements through collaboration, integration, and feedback, this article builds on the citizenship behaviors of employees and contributes originality by centering on the specific area of cognitive diversity and ramifications toward integration (National Intelligence Strategy 2019). This article addresses the question: *How can cognitive diversity increase integration endeavors across the 18 agencies to achieve the Intelligence Community's vision and support the mission*? The expectation for the IC to integrate also includes the workforce to be diverse and innovative. However, the review of the literature provides two major themes for group diversity. The first group diversity theme is positive as diversity increases innovation while the second is negative as group diversity tends to be divisive. The crux of the issue is that two of the DNI's main goals clash against each other due to the behavior of groups. This article will begin with a counterargument that will be nestled with cooperation theory to provide an argument that may seem counterintuitive. Although cooperation is perceived as a positive attribute, it actually presents a significant challenge for integration efforts of diverse populations. The next section will discuss cognitive diversity theory to frame the main argument on how cognitive diversity can enable integration. This study seeks to contribute to diversity theory by focusing on the ways cognitive diversity can increase integration within the IC. Finally, the article will conclude with implications for the IC to achieve its vision and mission focused on integration aspirations.

Although integration is a priority for senior leadership within the IC, there is a paucity of academic literature that focuses on the subject specifically linking avenues for this to occur. Therefore, this study directly contributes to the body of knowledge for IC integration endeavors. The research design selected to answer the central research question is a qualitative methodology with a case study that has positive association between crisis leadership and a cognitively diverse team. The findings from this exploratory social research may lead decision-makers to arrive at more informed decisions with less bias than just using personal experience or intuition.

Literature Review

The following literature review focuses on how the 18 agencies and organizations that comprise IC may become more integrated from three aspects. The article first examines cooperation theory to gain a deeper understanding of the criteria that enables and hinders cooperation in the work environment. Second, diversity theory is researched to understand the opportunities that diversity brings to the workforce. Finally, the literature review focuses on how cognitive diversity can bridge the gap that exists between strategy and execution of having a more integrated federation that is responsible for finding opportunities and thwarting threats that affect national security and national interests.

Cooperation Theory

Cooperation in the workplace focuses on employees working together to achieve organizational goals. Teams help fill ability and skill gaps experienced by individual employees by blending their differences and advising through their collective differences (Oosterhof, Van der Vegt, Van der Vliert, and Sanders 2009; Godneanu 2012). Individuals have different backgrounds, morals, values, and life experiences and therefore a certain degree of diversity occurs in the workplace. As a result of the added diversity, there is a changing of organizational dynamics that stem from the individual level yet may affect the entire community with positive and negative consequences. An example of a major bureaucratic reorganization is the Goldwater-Nichols Department of Defense Reorganization Act of 1986, which stated that the Secretary of each military department will cooperate fully with personnel of the Office of the Secretary of Defense (Public Law 99-433-Oct. 1, 1986). Although institutions may cooperate to integrate when ordered to, it may not be the optimal way to create buy-in. It should be noted that while the Goldwater Nichols Act is a great case to demonstrate integration within a government bureaucracy, it did take place several decades ago with different generations in the workforce. This point is provided to highlight current generations of employees have less institutional loyalty for long-term employment and thus key talent may be lost to turnover (Ertas 2015; Kritz 2018). It is essential for IC leaders and managers to understand and continue to work on retention efforts. The ODNI echoes this sentiment as they stated that due to complex national security threats, efforts must develop and retain its workforce that mirrors diversity (ODNI Diversity and Inclusion n.d.). Ertas' (2015) research focused on federal government employees, with an estimate that over 200,000 new employees would need to be hired to replace the retiring workforce. The literature depicts millennials as having less organizational loyalty than previous generations (Erta 2015). Previous research from Martin and Ottemann (2015) found the following work-related values for millennials:

> Ambitious to make a difference and secure a comfortable life, pro-work-life balance, satisfied with work tasks, interest in learning (fast, eager learners), desirous of security (not stability), collectivism, team player, optimistic, creativity (extremely expressive), unrealistic entitlement expectations, soft communication skills, value prompt recognition and reward, adaptable to new technologies, fun loving,

casual, socially conscious, multi-tasking is second nature, pro diversity (multi-cultural), self-confident, not easily intimidated (technically or interpersonally), expect instant gratification (impatient) (94).

Integration and diversity efforts within the IC are still a work in progress. From a historical perspective, Miner and Temes (2022) postulate "transforming diversity policy into action has remained consistently inconsistent" (21). As this phenomenon of diversity amongst employees brings both opportunities and challenges, two key questions emerge that senior organizational leaders may want to explore. The first question is to what extent will the workforce cooperate. The second question to ponder that may lead to future implications includes will the stakeholders and lower-level leaders seek to achieve institutional goals or have hidden agendas? For both questions, the answers may be illuminated or misguided by using rational choice as a framework. It may very well be a key assumption that employees will act rationally and cooperate during times of change and integration. To this end, "when a decision involves two or more interactive decision makers, each having only partial control over the outcomes, an individual may have no basis for rational choice without strong assumptions about how the other(s) will act" (Coleman 2003, 139).

Key assumptions to check when new organizations are integrated into a larger confederation include that the employees will get along with one another despite their differences. Regardless of the extent of diversity on a team, the beginning factor focuses on cooperation because, without cooperation, the institution's productivity suffers. Imagine a work environment where employees are brought together from different organizations and look at issues with different perspectives and are expected to integrate. Now multiply that by 18 and we then arrive at the focus of this article—integrating the 18 agencies and organizations that comprise the IC. The components of the IC work separately and together to conduct intelligence activities necessary for the conduct of foreign relations and the protection of national security (Office of the Director of National Intelligence 2021).

What happens if integration does not occur? If the institution happens to be the IC, an intelligence failure is more probable to occur. One piece of evidence comes from the past as Best (2011) states "by the two congressional intelligence committees and the 9/11 Commission (the National Commission on Terrorist Attacks Upon the United States) concluded that a central obstacle to acquiring advance information on the plot was the inability to bring together all information that had been acquired about the plotters—there were many clues[,] but they were retained in the files of different agencies" (5). One alternate future is that the U.S. may experience an attack similar to Pearl Harbor or the terrorist attacks of September 11, 2001. Unless we really understand the implications from failure and apply effective practices to mitigate future occurrences, then it is not a lesson learned, but an act that will repeat in a similar manner.

The charge to lead integration is truly an esemplastic effort. Leading the IC in intelligence integration is the ODNI's core mission. They construe intelligence integration as "synchronizing collection, analysis, and counterintelligence so that they are fused, effectively operating as one team" (ODNI How We Work n.d., n.p.). To create an environment of integration, individuals need to work together and cooperate effectively. A dictionary definition of cooperation includes "an act or instance of working or acting together for a common purpose or benefit; joint action" (Dictionary.com n.p., n.d.).

Researchers continue to explore cooperation in the work environment, and one major theme of the literature spotlights labor-management cooperation. Labor-management cooperation is essential to analyze as it focuses on decisions made between stakeholders within a more comprehensive institutional arrangement (Ospina and Yaroni 2003) and is applicable to leverage this framework to the IC as the IC is a large federation. The National Intelligence Strategy provides strategic direction across the IC bestowed by the DNI every four years. Integrated mission management is the first enterprise objective of the 2019 and most current National Intelligence Strategy. To achieve the objective, the first bullet within the National Intelligence Strategy states the IC will "Provide leadership and community management to foster collaboration, streamline processes, and effectively manage resources to achieve IC mission objectives" (18). Collaboration is a barrier within the IC for several reasons. The first reason includes an archaic Cold War culture to willfully not share intelligence with different organizational or agency intelligence professionals (Hamrah 2013). A second reason comes from Miles (2015) who adroitly stated, "functional and geographic separation can lead to or exacerbate differences in organizational culture ... that impede the desire to collaborate" (317).

Former President Obama issued the 2011 executive order 13583—Establishing a Coordinated Government-wide Initiative to Promote Diversity and Inclusion in the Federal Workforce. As the goal of the DNI is greater integration, then this author argues that Krtiz's "Model of Integration" [Communication x (Cooperation x Collaboration) = Integration] is a path to achieve the goal. Collaboration cannot be achieved without cooperation nor communication. For labor-management cooperation to occur, it takes a decision at the individual level to partake in cooperative behavior (Ospina and Yaroni 2003). Cooperation is both an attitude and an action. Cooperation Theory explains that employees swim or sink together due to the perception of goals being linked and thus tend to cooperate rather than compete (Godeanu 2012). Employees can be motivated to engage in cooperative behavior when both sides conclude that the incentive to cooperate yields a greater benefit from joint efforts if the act occurs (Cooke 1990; Ospina and Yaroni 2003).

Cognitive Diversity

The study of diversity and how diversity affects group dynamics, organizational culture, and organizational goals is a worthy pursuit in research. Although new knowledge is gained when diversity is researched, there is much to contribute toward its theory and as Austin (1997) argued the overall understanding remains fragmented. Diversity is important because it is a primer for innovation. The DNI recognizes the importance of diversity not only to create innovation but to build and retain a highly-skilled workforce (ODNI 2019). While there are numerous definitions for diversity in the literature, it is fitting to select the one most relevant to the topic. The DNI defines diversity as "a collection of individual attributes that together help IC elements pursue organizational objectives efficiently and effectively" (ODNI 2019, 20).

Not all diversity is beneficial to achieve organizational goals, and therefore the types of diversity matter for teambuilding. As a simple example, if a child were to take a peanut butter and jelly sandwich and pour orange juice on it, the ingredients would undoubtedly be diverse, but the result would be a mess. The same holds with individuals. Researchers Gyngell and Easteal (2015) presented the argument "aspects of our moral psychology make it difficult for people to cooperate and coordinate actions with those who are very different from themselves" (66). Although our similarities may be closer than our collective differences, we are on different locations of the diversity spectrum. Moreover, because of this diversity, cooperation may be more challenging to achieve for many reasons. Challenges with cooperation may emerge that affect communication and coordination with members of diverse teams (Gyngell and Easteal 2015).

It would be hard to argue that there was a time when organizations were more heterogenous than today. Academic strategic alliances, business mergers, and changes to the U.S. IC's organizational structures are a few examples. Diversity will continue to increase amongst institutions, and this is a beneficial phenomenon for stakeholders. Previous research established that heterogeneous teams comprised of low and high skilled employees outperform homogenous teams regardless of ability (Hamilton, Nickerson, and Owan 2003; Godeanu 2012). Although race, gender, and age certainly contribute to diversity, these types align more with a diversity of demographics. This research moves the argument into a deeper consideration of diversity that focuses on the diverse ways employees think about issues. This is important because groupthink while having high levels of cooperation, may lead to a low and dangerous level of decision-making. As Lowenthal (2015) argues:

> The United States developed the concept of competitive analysis, an idea that is based on the belief that by having analysts in several agencies with different backgrounds and perspectives work on the same issue, parochial views more likely will be

> countered ... and proximate reality is more likely to be achieved ... [and] be an antidote to groupthink and forced consensus (17).

Callum (2001) argued the following regarding intelligence analysis and homogeneity as it "dooms the community to what is essentially one point of view on complex issues, and makes it all but certain that in the chaotic and uncertain post-Cold War international arena, U.S. intelligence will be fated to failures of greater number and magnitude" (26).

There are numerous definitions of cognitive diversity found within the literature. Some of the definitions were quite limited such as O'Donovan's (2010) writing "the diversity of cognition generated by cognitive disabilities" (172) as cognitive diversity is far more comprehensive than those who have cognitive disabilities. A more robust yet still abridged definition of cognitive diversity is defined "as variation in underlying attitudes, beliefs or values developed through individual experience and background" (Tegarden, Tegarden, and Sheetz 2009, 538). Within this study, this researcher defines cognitive diversity as the variation of thoughts amongst a team that comes through each individual's attitudes, beliefs, career backgrounds, cultural experiences, educational backgrounds, language abilities, locational backgrounds, morals, philosophical perspectives, personal life experiences, socioeconomic backgrounds, and values. There are numerous variables in my definition because people are complex, and adaptiveness continuously emerges. Dr. Scott E. Page (2011), an American social scientist and John Seely Brown Distinguished University Professor of Complexity, asserts "systems that produce complexity consist of diverse rule-following entities whose behaviors are interdependent .... In addition, the entities often adapt. That adaptation can be learning in a social system" (17).

Although cognitive diversity may not arrive at the solution immediately, it will provide context and deeper thinking for decision-makers to select better decisions and arrive at a beneficial solution quicker than a team with less cognitive diversity. The Central Intelligence Agency (CIA) states they are mission dependent on diversity and inclusion as they need their officers to provide "a full range of perspectives, experiences, and talents to our mission, the Agency will be better prepared to address intelligence challenges and support its customers" (Working at CIA n.d., n.p.). Cognitive diversity positively affects team performance, notably during the beginning stages of strategic planning (Tegarden, Tegarden, and Sheetz 2009); this is why cognitive diversity is an important aspect for the IC. Further, previous research in business, economics, and other social sciences demonstrates that groups with more diversity outperform groups with less diversity when dealing with complexity (Gyngell and Easteal 2015). This understanding of knowledge is essential for the IC as its prominent role is finding opportunities for and mitigating threats against national security endeavors. It is an understatement to write that the IC faces complexity with every challenge as the goal of intelligence is to find truth, and denial and deception campaigns of adversaries often obscure the truth.

Organizational culture is a consideration for leading to a diverse workforce and then enablement of integration. Employees are often uncomfortable about change unless they perceive that the change will positively affect them. One such occurred when at the time, James Clapper, newly appointed DIA Director, who faced an employee revolt that caused him to quickly realize the difference in mindsets between military members who are accustomed to moving frequently within organizations and to new physical locations contrasted with a civilian workforce who are accustomed to staying within the same seat of an organization for multiple years. Clapper (2018) stated "whatever merits our reorganization plan had, we hadn't sufficiently talked with the employees and worked out the details ... I learned that gaining employees' buy-in before making big changes is essential, and that there are cultural differences between military and civilian employees" (73). Change in culture is difficult to manage in one organization. The degree of difficulty increases as the size of integration efforts increase. This is especially true when new or outside agencies are brought into the fold of a federation, as the culture can either help or hinder integration depending on workplace attitude and moral psychology.

As this article specifically focuses on the IC, there was a change to the federation in the first month of 2021 as the United States Space Force was officially included as the 18th agency. During these periods of change, employees have new opportunities to work with employees from different agencies. The former National Geospatial-Intelligence Agency Director, Robert Cardillo (2010), argues that analytic tradecraft is the IC's culture. One scholar of organizational culture, Schein (2004), as referenced in Cardillo (2010), defined organizational culture as:

> A pattern of basic assumptions—invented, discovered, or developed by a given group as it learns to cope with its problems of external adaption and internal integration—that has worked well enough to be considered valid, and therefore, to be taught to new members as the correct way to perceive, think, and feel in relation to those problems (1).

Research Methods

While the DNI wants IC integration to occur, part of the solution lies with practitioners responsible for doctrine and the other half with academics to develop theory. Barger (2004) presented a strong argument "as no theory or doctrine exists to explain the changing role of intelligence in a changing world, the Intelligence Community also lacks an overarching strategy to meet new security challenges" (24). The first DNI, John Negroponte, countered this argument one year later by presenting the National Intelligence Strategy of the United States of America: Transformation through Integration and Innovation. Within this document, the DNI acknowledged former U.S. President Bush 43's task to "integrate the domestic and foreign dimensions of US intelligence so that there are no gaps in our understanding of threats to our national security" (ODNI 2005, 3). In a letter signed within the 2019 National Intelligence Strategy by the previous DNI, Daniel Coats, his intentions to the IC are evident as the first two of four bullet statements depict:

- Increase integration and coordination of our intelligence activities to achieve best effects and value in increasing our mission
- Bolster innovation to constantly improve our work (5).

Models can enable decision-makers to refine thinking through broad application for their usage to include how countries determine national security structures, including assessments to basic decision-making and policy, priorities, and resources (Kritz 2021, 105). Tuckman's 1965 influential Developmental Sequence model hypothesized groups moving throughout the following stages forming, storming, norming, performing and adjourning (Tuckman and Jensen 2010). This model was considered as a framework to aid the IC toward integration across the federation however Tuckman's model is focused on small groups (15-20 individuals) and the IC is comprised of over tens of thousands of employees. Therefore, a larger model is needed to properly scope integration endeavors. One model that adds insight comes from the research of Miczka and Größler (2010) that focused on the dynamics evolving from two converging organizational cultures and changes in employee commitment. The "Merger dynamics" model depicted "the number of interactions and the willingness to cooperate" as the two emerging variables (Miczka and Größler 2010, 1501).

This article uses a qualitative methodology with a case study as the research method. The ways scientists study and bestow explanations through scientific inquiry are diverse. Although the case selected comes from the medical profession, case studies regardless of career sector may be useful as they provide a framework for analysis and help describe how solutions may be adapted within similar solutions, specifically for this article, the IC in times of crisis. Further, as a former program director, assistant professor, and intelligence officer at the National Intelligence University, this researcher heard on numerous occasions from senior leaders at the director level that the IC needs to leverage the business sector and others to help with innovation. Therefore, it seemed fitting to select a case the IC can leverage that occurred outside the IC. The research is focused on group behavior and how individuals within distinct groups can integrate under the IC federation to accomplish the DNI's vision and mission. To that end, a qualitative methodology seemed to be the most fitting to answer the central research question. Creswell (2009) defines qualitative research as "a means for exploring and understanding the meaning individuals or groups ascribe to a social or human problem ... the final written report has a flexible writing structure" (232). A case study was selected as the research method to answer the central research question: *How can cognitive diversity increase integration endeavors across the 18 agencies to achieve the Intelligence Community's vision and support the mission?* Creswell (2009) defines case studies as "a qualitative strategy in which the researcher explores in depth a program, event, activity, process, or one or more individuals" (227). Ragin (2009) states, "implicit in most social scientific notions of case analysis is the idea that the objects of investigation are similar enough and separate enough to permit treating them as comparable instances of the same general phenomenon" (1). Further, "the idea of comparable cases is implicated in the boundary between dominant forms of social science and other types of discourse about social life" (Ragin 2009, 2). It should be noted that case studies are not bound by nationality. The findings from this exploratory social research may lead decision-makers to arrive at more informed decisions with less bias than just using personal experience or intuition.

The Case Study: Cognitive Diversity as the Quality of Leadership in Crisis: Team Performance in Health Service during the COVID-19 Pandemic

Research conducted by Joniaková, Jankelová, Blštáková, and Nèmethov (2021) was selected as a case study as the authors "justified the positive association between crisis leadership and team performance, which is mediated by cognitive diversity, supporting the quality of decision-making in crisis leadership" (Joniaková et al., 2021, 1). The case study is relevant and relatable to the IC as the DNI's third objective of the National Intelligence Strategy is that people and leadership will meet the objective by shaping "a diverse workforce with the skills and capabilities needed to address enduring and emerging requirements" (20). Using Joniaková et al.'s (2021) research as a case study to explore their event on cognitive diversity is also relevant to answer the central question and for the IC, as the DNI's fourth objective is innovation and argues that the IC must "foster unconventional thinking and experimentation that addresses new, better ways of accomplishing the IC's mission, especially those approaches that emphasize acceleration, simplicity, and efficiency without sacrificing quality and outcomes" (National Intelligence Strategy 2019, 21).

Joniaková et al.s' (2021) research explored the following problem: leaders of healthcare providers are challenged with managing teams of healthcare professionals during crises. Their study suggests that human behavior is the primary variable to determine organizational efficiency. The human element is not confined to this case but expands to all organizations, including the IC. Joniaková et al. (2021) study focused on crisis leadership (CL), team performance, cognitive diversity, and decision-making during a crisis. Callum (2001) makes a compelling argument on the human element in what he labels as the fundamental weakness of the IC and why cognitive diversity is essential for the IC. He states, "with many intelligence professionals cut from the same cultural cloth, analysts share 'unacknowledged biases' that circumscribe both the definition of problems and the search for solutions" (26). Joniaková et al. (2021) argue that uncertainty creates barriers to employees during a crisis. In comparison the IC often works within periods of crises that are often labeled as intelligence failures. Examples include Pearl Harbor, the Cuban Missile Crisis, and 9/11. The most common leadership themes Joniaková et al. (2021) determined within their literature review with a baseline variable they referred to as CL. Their research depicted four hypotheses: Hypothesis 1. CL is positively associated with medical teams' performance (MTP); Hypothesis 2. The relationship between CL and MTP is positively mediated by crisis decision-making (CDM); Hypothesis 3. The relationship between CL and MTP is positively mediated by CDM; and Hypothesis 4. The relationship between CL and MTP is sequentially and positively mediated by Cognitive Diversity (CDL) and CDM.

Joniaková et al. (2021) used a questionnaire as a qualitative research tool that was electronically sent to healthcare facilities in Slovakia. Their research sample was 216 leaders (team leaders of medical teams) of private and public hospitals following the COVID-19 outbreak in Slovakia. Joniaková et al. (2021) control variables included "size of the healthcare facility according to the number of employees, gender and age of the team leader, his position in the managerial hierarchy, and the length of experience in the team leader position" (6).

Case Study Conclusion

Joniaková et al. (2021) depicted relationships between individual variables determined by means of a correlation matrix. Of importance, "the hypothesis of dependence between leadership competencies and team performance, which is mediated by the quality of decision-making, supported by the cognitive diversity, has been confirmed by the research" (11).

Analysis

From the 2022 Annual Threat Assessment of the U.S. IC, infectious diseases and the impact of the COVID-19 pandemic is one of the most serious threats facing the U.S. While IC professionals are certainly not healthcare professionals, the DNI is responsible for integrating teams of IC professionals in the form of agencies and organizations that comprise the IC federation during times of peace and crisis. The challenges facing the DNI are complex and challenging as having a cognitively diverse workforce enhances one issue (innovation) and yet inhibits another (cooperation). Austin (1997) summed up the literature's findings that demonstrates evidence of opposing outcomes by stating "the two most consistent findings in the group diversity literature are that increased group diversity lead to (1) an increase in innovation and thinking, and (2) a decrease in group cohesion, and subsequent increase in intra-group conflict" (348). The results from the case study found a positive outcome with cognitive diversity. This has implications for the IC that can lead to a path toward integration. If the IC is slow to select a cognitively diverse workforce, then the DNI's mission and vision may not be achieved effectively nor efficiently. Hopefully, it will not take another catastrophic event similar to 9/11 to drive monumental change.

Within the second objective of the 2019 National Intelligence Strategy, the DNI tasked the IC to "promote and identify best business practices and functions to optimize solutions and increase collaboration to create a culture of continuous learning, innovation, and partnerships across the community" (19). To aid task fulfillment, this author noticed four main themes from Joniaková et al.'s (2021) research that IC decisionmakers can apply toward integration efforts: 1) Speed of leadership decision-making; 2) Cognitive diversity of teams managing crises; 3) Leadership sets example during times of crises; and 4) Promoting a culture of trust and teamwork.

Speed of Leadership-Decision Making

Timeliness is an essential aspect of intelligence. Getting the correct information to the right people at the right time is paramount and can be the difference between an issue becoming a policy success or an intelligence failure. Intelligence follows a five-step process: 1) Planning and direction; 2) Collection; 3) Processing; 4) Analysis and production; and 5) Dissemination. One can observe how the timeliness of leadership-decision making can affect all of the steps; however, the most felicitous step to focus upon is planning and direction. Planning and direction begin and concludes with leadership. To date, there have been seven DNIs elected to fulfill the position. A powerful start came from the fourth and longest serving DNI, James Clapper, who while describing the new IC IT Enterprise (IC ITE sounds like Eye Sight) from concept to implementation was instrumental in increasing intelligence integration and driving efficiencies (IC ITE Strategy 2016 – 2020). This sentiment expands on how intelligence is conducted to how teamwork and integration are essential to conduct its mission. As stated by the Office of the DNI (ODNI) regarding its core mission:

> To lead the IC in intelligence integration, forging a community that delivers the most insightful intelligence possible. That means effectively operating as one team: synchronizing collection, analysis and counterintelligence so that they are fused. This integration is the key to ensuring national policymakers receive timely and accurate analysis from the IC to make educated decisions.

Cognitive Diversity of Teams Managing Crises

Cognitive diversity is essential because senior leaders within the IC are looking to younger generations of the workforce to be innovative. O'Donovan (2010) stated "as one aspect of intellectual diversity, cognitive diversity promises novel ways of thinking and new understandings of what knowledge is, who makes it, and how it is made" (172). Innovation is important to the IC and, as the fourth objective in the DNI's National Intelligence Strategy "is critical to ensuring that the IC can provide the strategic and tactical decision advantage that policymakers and warfighters require" (21). Further, it takes innovation to work through complexity and mitigate wicked issues. Wicked issues are considered as such due to differences in priorities amongst stakeholders and as they cannot be solved by a single organization or sector (Sachs, Rühli, and Meier 2010). Integration and innovation increase when teams can interact with those who have the least in common, and the empathy gap is mitigated. Diversity enables innovation by applying differences in abilities, experiences, and knowledge (Andresen 2007). The probability of achieving a high-quality solution increases with the level of group creativity when a team has a greater variety of perspectives on a problem-set (Hoffman 1979; Austin 1997).

Having a team of cognitively diverse employees who can cooperate mitigates groupthink by having rigorous discussions about the teamwork by their individual mental frameworks that, when openly discussed, creates a refinement of thought for the entire team. Previous research found that when diversity in teams is increased, it often causes an increase in knowledge production, and solving complex issues is resolved through various background beliefs, concepts used, heuristics, reasoning styles, and representations (Pöyhönen 2017; Page 2019). The interaction of individuals with differing backgrounds increases opportunities to learn from each other. This is important because when cognitively diverse individuals interact and discuss an issue, a solution toward that issue may not immediately emerge; however, refinement of thought and deeper understanding of the issue occurs between the individuals, and with enough encounters, the organizational culture changes.

Leadership Setting the Example During Times of Crises

To achieve success for the DNI and answer the central question for this article, IC leaders from the top down will need to work with employees within their own organizations and agencies and the other IC agencies. This will take time as trust is the currency of leadership to change bureaucratic culture slowly. An emphasis on cognitive diversity as an organizational priority takes collective buy-in. A change in organizational culture may be both difficult and take a long period of time. In short, leaders, managers, and employees will need to demonstrate a desire to practice continuous learning. Further, effective communication, focused programs, and a high degree of trust between employees and bosses are factors that can positively affect organizational change.

Promoting a Culture of Trust and Teamwork

Trust issues within the IC stem from the lack of inter-agency collaboration and information gaps caused by a lack of sharing (Segall 2010). To promote a culture of trust and teamwork, director level IC leadership should continue to message the workforce the reasons why integration is vital to achieving the IC's vision and mission while simultaneously the federation must have a willingness to cooperate and collaborate with one another. The first DNI stated when discussing national intelligence that "it must recognize that its various institutional cultures developed as they did for good reasons ... that all cultures evolve or expire, and the time has come for our domestic and foreign intelligence cultures to grow stronger by growing together" (ODNI 2005, 3). This message should be presented to the IC workforce continuously. One method to achieve this endeavor is to continue to build upon the sense of community within the IC.

There are genuinely unique hurdles that face federal entities, including the IC. To become a member of the IC, a security clearance is needed due to the nature of working with classified information. Kyzer (2021) reports that the average processing time in 2021 for a Secret clearance is 132 days and 159 days for a Top Secret clearance. One of the implications with the time constraint is that it demonstrates cognitive diversity let alone normal hiring procedures for the IC does not occur instantaneously. The process of hiring individuals who can obtain a security clearance and pass a polygraph test if required is a stark reminder that security comes before diversity. While security requirements for hiring is an opportunity to think through with hiring talent who may not wait over five months for a change in career paths.

Concluding Thoughts

This article commenced with an argument that the IC needs to make fundamental changes to remain relevant. The former DNI, James Clapper, correctly stated that IC must think differently to support customers as the threats continue to evolve (IC ITE Strategy 2016–2020). Two questions that the IC agency directors should ask include, what integration looks like and what are the expectations. While previous researchers such as Callum (2001) proposed a solution to imbue the IC with a competitive array cultures, mindsets, and ideas by expanding diversity, there is a challenge to overcome as found within the literature, diversity can act as an inhibitor to cooperation. Thus, leadership needs to be proactive with leading the workforce to provide a work environment that welcomes openness to diverse thoughts. Evidence to support this recommendation comes from the four themes from the discussion section of Joniaková et al.'s (2021) case study that IC decisionmakers can apply toward integration efforts: 1) Speed of leadership decision-making; 2) Cognitive diversity of teams managing crises; 3) Leadership sets example during times of crises;

and 4) Promoting a culture of trust and teamwork.

Having employees with different skillsets matters when it comes to working through complexity or mitigating wicked issues. The right mixture of individuals should be taken into account when forming teams, as not all diversity yields positive results when it comes to building teams for the purpose of innovation. Miner and Temes (2022) argue the importance of increasing diversity within the IC as they warn that "if the IC does not continue to routinely assess barriers to its personnel policies, practices, and programs, it will limit itself to its old ways of recruitment, hiring, retention, and promotion and will miss out on the best talent, perspectives, and capabilities" (24). Characteristics worthy of consideration include selecting individuals with different life experiences including educational backgrounds and professional experiences. Teams that are cognitively diverse are able to problem-solve more effectively through refined thought, not as susceptible to groupthink, and create multiple solutions. Finally, the application of Kritz's "Model of Integration" [Communication x (Cooperation x Collaboration) = Integration] may help the DNI achieve greater integration across the IC.

Bibliography

Andresen, Maike. "Diversity Learning, Knowledge Diversity and Inclusion: Theory and Practice as Exemplified by Corporate Universities." *Equal Opportunities International 26,* no. 8 (2007): 743-760. DOI:10.1108/02610150710836118. https://www.proquest.com/scholarly-journals/diversity-learning-knowledge-inclusion/docview/199673493/se-2?accountid=10504.

Ateş, Ahmet. "The Transformation of Russian Intelligence Community After the Cold War (1991-1993)." *Journal of Black Sea Studies Karadeniz Arastirmalari*, no. 66 (Summer, 2020): 321-32, https://www.proquest.com/scholarly-journals/transformation-russian-intelligence-community/docview/2425600373/se-2?accountid=10504.

Austin, John R. "A Cognitive Framework for Understanding Demographic Influences in Groups." *International Journal of Organizational Analysis 5,* no. 4 (10, 1997): 342-59, https://www.proquest.com/scholarly-journals/cognitive-framework-understanding-demographic/docview/198653645/se-2?accountid=10504.

Barger, Deborah G. "It is Time to Transform, Not Reform, U.S. Intelligence." *The SAIS Review of International Affairs 24,* no. 1 (Winter, 2004): 23-31, https://www.proquest.com/scholarly-journals/is-time-transform-not-reform-u-s-intelligence/docview/231317038/se-2?accountid=10504.

Best, Richard A. "Intelligence Information: Need-to-Know Vs. Need-to-Share." *Federation of American Scientists,* 2011. https://www.proquest.com/reports/intelligence-information-need-know-vs-share/docview/1820857175/se-2?accountid=10504.

Cardillo, Robert. "Intelligence Community Reform: A Cultural Evolution." *Studies in Intelligence 54,* no. 3 (Extracts, September 2010).

Clapper, James R.; Brown, Trey. *Facts and Fears: Hard Truths from a Life in Intelligence,* 2018. Penguin Publishing Group. Kindle Edition. (73).

Colman, Andrew M. "Cooperation, Psychological Game Theory, and Limitations of Rationality in Social Interaction." *Behavioral and Brain Sciences 26, no. 2* (04, 2003): 139-53, https://www.proquest.com/scholarly-journals/cooperation-psychological-game-theory-limitations/docview/2027511669/se-2?accountid=10504.

Cooke, William N. 1990. *Labor Management Cooperation: New Partnership or Going in Circles?* Kalamazoo, MI: Upjohn Institute for Employment Research.

Dictionary.com. Cooperation. Accessed on 14 April 2022. https://www.dictionary.com/browse/cooperation

Ertas, Nevbahar. 2015. "Turnover Intentions and Work Motivations of Millennial Employees in Federal Service." *Public Personnel Management 44,* no. 3: 401-423, https://www.proquest.com/scholarly-journals/turnover-intentions-work-motivations-millennial/docview/1761254511/se-2?accountid=8289.

Godeanu, Ana-Maria. "The Determinants of Helping Behavior in Teams." *Management & Marketing 7,* no. 3 (2012): 393-414, https://www.proquest.com/scholarly-journals/determinants-helping-behavior-teams/docview/1036593352/se-2?accountid=10504.

Gyngell, Chris and Simon Easteal. 2015. "Cognitive Diversity and Moral Enhancement: CQ." *Cambridge Quarterly of Healthcare Ethics 24,* no. 1: 66-74, https://www.proquest.com/scholarly-journals/cognitive-diversity-moral-enhancement/docview/1664744746/se-2?accountid=10504.

Hamilton, Barton H., Jack A. Nickerson, and Hideo Owan. "Team Incentives and Worker Heterogeneity: An Empirical Analysis of the Impact of Teams on Productivity and Participation." *The Journal of Political Economy 111,* no. 3 (06, 2003): 465-97, https://www.proquest.com/scholarly-journals/team-incentives-worker-heterogeneity-empirical/docview/195425614/se-2?accountid=10504.

Hamrah, Satgin S. "The Role of Culture in Intelligence Reform." *Journal of Strategic Security* 6, no. 3 (2013): 160-71, https://www.proquest.com/scholarly-journals/role-culture-intelligence-reform/docview/2205364334/se-2?accountid=10504.

IC ITE Strategy 2016–2020. Accessed on 8 August 2021. https://www.dni.gov/files/documents/CIO/IC%20ITE%20Strategy%202016-2020.pdf

Joniaková, Zuzana, Nadežda Jankelová, Jana Blštáková and Ildikó Némethová. "Cognitive Diversity as the Quality of Leadership in Crisis: Team Performance in Health Service during the COVID-19 Pandemic." *Healthcare* 9, no. 3 (2021): 313, https://www.proquest.com/scholarly-journals/cognitive-diversity-as-quality-leadership-crisis/docview/2501890794/se-2?accountid=10504.

Kritz, David J. "A Model Approach: Considering Models to Enhance Intelligence Utility for National Security Decision-Making." *American Intelligence Journal (38)*1, Spring 2021

Kritz, David J. (2018). "Why Ethical Leadership Matters: A Case Study to Improve Military Specialists' Employee Retention Rates." *The International Journal of Ethical Leadership: Vol. 5*, Article 17.

Kyzer, Lindy. "How Long Does It Take to Process a Security Clearance—Q2 2021 Update." *Clearance Jobs* Apr 14, 2021. Accessed on 24 August 2021. https://news.clearancejobs.com/2021/04/14/how-long-does-it-take-to-process-a-security-clearance-q2-2021-update/

Lowenthal, Mark, M. *Intelligence: From Secrets to Policy* (California: CQ Press, 2015), 17.

Martin, Thomas N. and Robert Ottemann. 2015. "Generational Workforce Demographic Trends and Total Organizational Rewards Which Might Attract and Retain Different Generational Employees." *Journal of Behavioral and Applied Management 16*, no. 2: 91-115, https://www.proquest.com/scholarly-journals/generational-workforce-demographic-trends-total/docview/1891260662/se-2?accountid=8289.

Miczka, Switbert and Andreas Größler. "Merger Dynamics: Using System Dynamics for the Conceptual Integration of a Fragmented Knowledge Base." *Kybernetes 39,* no. 9 (2010): 1491-1512. DOI:10.1108/03684921011081132. https://www.proquest.com/scholarly-journals/merger-dynamics/docview/761430599/se-2?accountid=10504.

Miles, Anne, D. "Intelligence Education and Integration: A Symbiotic Relationship," in *Intelligence Management in the Americas,* eds. Russell G. Swenson and Carolina Sancho Hirane (Washington D.C.: NI Press, 2015), 317-320.

Miner, Michael and Lindsay Temes. "The Past, Present, and Future of Diversity, Equity, and Inclusion in the American Intelligence Community." Harvard Kennedy School Belfer Center for Science and International Affairs Report April 2022.

The National Commission on Terrorist Attacks Upon the United States. 2004. *The 9/11 Commission Report.* Accessed on 20 July 2021. https://9-11commission.gov/report/

O'Donovan, Maeve, M. "Cognitive Diversity in the Global Academy: Why the Voices of Persons with Cognitive Disabilities are Vital to Intellectual Diversity." *Journal of Academic Ethics 8,* no. 3 (09, 2010): 171-85, https://www.proquest.com/scholarly-journals/cognitive-diversity-global-academy-why-voices/docview/816534665/se-2?accountid=10504.

Office of the Director of National Intelligence. Diversity and Inclusion. Accessed on 14 April 2022. https://www.dni.gov/index.php/how-we-work/diversity

Office of the Director of National Intelligence. How We Work. Accessed on 14 April 2022. https://www.dni.gov/index.php/how-we-work

Office of the Director of National Intelligence (2019). The National Intelligence Strategy of the United States of America 2019. Accessed on 28 June 2021. https://www.dni.gov/index.php/newsroom/reports-publications/item/1943-2019-national-intelligence-strategy

Office of the Director of National Intelligence (2005). The National Intelligence Strategy of the United States of America: Transformation through Integration and Innovation 2005. Accessed on 21 August 2021. https://www.dni.gov/files/documents/Newsroom/Reports%20and%20Pubs/NISOctober2005.pdf

Office of the Director of National Intelligence. What is Intelligence? Accessed on 26 June 2021. https://www.dni.gov/index.php/what-we-do/what-is-intelligence

Oosterhof, Aad, Gerben Van der Vegt S., Evert Van de Vliert, and Karin Sanders. "Valuing Skill Differences: Perceived Skill Complementarity and Dyadic Helping Behavior in Teams." *Group & Organization Management 34,* no. 5 (10, 2009): 536-62, https://www.proquest.com/scholarly-journals/valuing-skill-differences/docview/1928228145/se-2?accountid=10504.

Ospina, Sonia and Allan Yaroni. 2003. “Understanding cooperative behavior in labor management cooperation: A theory-building exercise.” *Public Administration Review 63,* no. 4: 455-471, https://www.proquest.com/scholarly-journals/understanding-cooperative-behavior-labor/docview/197175643/se-2?accountid=10504.

Page, Scott E. Diversity and Complexity. (p. 17). Princeton University Press. (2011) Kindle Edition.

Page, Scott E. The Diversity Bonus: How Great Teams Pay Off in the Knowledge Economy. Princeton University Press. (2019). Kindle Edition.

Pöyhönen, Samuli. 2017. Value of cognitive diversity in science.” *Synthese 194,* no. 11: 4519-4540, https://www.proquest.com/scholarly-journals/value-cognitive-diversity-science/docview/1963841905/se-2.

Public Law 99-433-Oct. 1, 1986. “Goldwater–Nichols Department of Defense Reorganization Act of 1986.” Accessed 3 July 2021. https://history.defense.gov/Portals/70/Documents/dod_reforms/Goldwater-NicholsDoDReordAct1986.pdf

Ragin, Charles. What Is a Case? (p. 1). Cambridge University Press. (2009). Kindle Edition.

Rhodes, Richard. (2011) Book Review of “Top Secret America: The Rise of the New American Security State” by Dana Priest and William M. Arkin. Accessed August 1, 2021. https://www.washingtonpost.com/entertainment/books/top-secret-america-the-rise-of-the-new-american-security-state-by-dana-priest-and-william-m-arkin/2011/09/30/gIQAvkkUkL_story.html

Simon, Ruth. “Covid-19’s Toll on U.S. Business. 200,000 Extra Closures in Pandemic’s First Year.” *The Wall Street Journal* April 16, 2021. Accessed 13 April 2022. https://www.wsj.com/articles/covid-19s-toll-on-u-s-business-200-000-extra-closures-in-pandemics-first-year-11618580619

Tegarden, David P., Linda F. Tegarden, and Steven D. Sheetz. “Cognitive Factions in a Top Management Team: Surfacing and Analyzing Cognitive Diversity using Causal Maps.” *Group Decision and Negotiation 18,* no. 6 (11, 2009): 537-66, https://www.proquest.com/scholarly-journals/cognitive-factions-top-management-team-surfacing/docview/223822109/se-2?accountid=10504.

Tuckman, Bruce W. and Mary Ann C. Jensen. “Stages of Small-Group Development Revisited.” *Group Facilitation, no. 10* (2010): 43-8, https://www.proquest.com

/scholarly-journals/stages-small-group-development-revisited1/docview/747969212/se-2?accountid=10504.

Tüzüner, Musa. "Insights of Intelligence Insiders on (Non-) Sharing Intelligence Behaviors." *All Azimuth 3,* no. 2 (07, 2014): 51-66, https://www.proquest.com/scholarly-journals/insights-intelligence-insiders-on-non-sharing/docview/1658889134/se-2?accountid=10504.

Working at CIA—Diversity and Inclusion. Accessed on 14 April 2022. https://www.cia.gov/careers/working-at-cia/diversity/

David J. Kritz holds a Doctorate in Business Administration and an MS in International Relations. He is an Associate Professor of Intelligence Studies in the school of Security and Global Studies for the American Military University, an adjunct professor for SUNY Empire State College, and an official reviewer for the *Journal of Leadership Education.* He recently retired from the U.S. Air Force as an intelligence officer and the key position of MSSI Program Director, and an Assistant Professor at the National Intelligence University. His research emphasis focuses on the U.S. Intelligence Community, national security issues, and complexity.

Appendix A

The below chart represents the 18 agencies and organizations, including the ODNI that comprise the IC led by the DNI.

Retrieved from: https://www.dni.gov/index.php/what-we-do

Physician, then Political Dictator: Bashar al-Assad, President of the Syrian Arab Republic

Casey Skvorc and Nicole K. Drumhiller

Abstract

This article examines the life influences and circumstances that enabled President Bashar al-Assad to shift from the practice of medicine to becoming a brutal political dictator. Assad is one among the few members of the medical community who have transitioned from doctor to political dictator. Widely known for his deadly rule over Syria, his formal education and training were in medicine, specializing in ophthalmology. He graduated from Damascus Medical University, briefly practiced in a Syrian military hospital, and later sought to advance his training at the Western Eye Hospital in London. Following his father's death, Bashar al-Assad assumed power and has ruled Syria as its president for over 20 years. Using psychobiography, this case study assesses Bashar al-Assad's brutal transformation from doctor to dictator. His leadership sophistication and social elevation were originally enhanced from his upbringing as the second son of the President of Syria and his training as a physician. While he initially chose to specialize in Ophthalmology, reportedly because of the reduced exposure to blood, al-Assad is responsible for the deaths and disappearance of approximately 500,000 people.

Keywords: psychobiography, leadership analysis, doctator, dictator, authoritarian, at-a-distance-assessment

Médico, entonces dictador político: Bashar al-Assad, presidente de la República Árabe Siria

Resumen

Este artículo examina las influencias de la vida y las circunstancias que permitieron al presidente Bashar al-Assad pasar de la práctica de la medicina a convertirse en un brutal dictador político. Assad es uno de los pocos miembros de la comunidad médica que ha pasado de médico a dictador político. Ampliamente conocido por su gobierno mortal sobre Siria, su educación y formación formales fueron en medicina, especializándose en oftalmología. Se graduó de la

 doi: 10.18278/gsis.7.1.4

Universidad de Medicina de Damasco, practicó brevemente en un hospital militar sirio y luego buscó avanzar en su formación en el Western Eye Hospital de Londres. Tras la muerte de su padre, Bashar al-Assad asumió el poder y ha gobernado Siria como su presidente durante más de 20 años. Utilizando la psicobiografía, este estudio de caso evalúa la brutal transformación de Bashar al-Assad de médico a dictador. Su sofisticación de liderazgo y elevación social se mejoraron originalmente a partir de su educación como segundo hijo del presidente de Siria y su formación como médico. Si bien inicialmente eligió especializarse en Oftalmología, al parecer debido a la menor exposición a la sangre, al-Assad es responsable de la muerte y desaparición de aproximadamente 500.000 personas.

Palabras clave: psicobiografía, análisis de liderazgo, doctor, dictador, autoritario, evaluación a distancia

从医生到政治独裁者：阿拉伯叙利亚共和国总统巴沙尔·阿萨德

摘要

本文分析了使巴沙尔·阿萨德总统从行医转变为残酷的政治独裁者的一系列生活影响和环境。阿萨德是医学界为数不多的、从医生转变为政治独裁者的成员之一。他对叙利亚的极端统治广为人知，但他接受的正规教育和培训是医学（专攻眼科）。他毕业于大马士革大学医学系，曾在叙利亚军队医院短暂行医，之后试图在伦敦西部眼科医院进修培训。在其父去世后，巴沙尔·阿萨德掌权并作为总统统治叙利亚超过20年。本案例研究使用心理传记学，评估巴沙尔·阿萨德从医生到独裁者的残酷转变。他的领导才能和社会地位最初是从他作为叙利亚总统的第二个儿子的成长经历以及他的内科医生培训中得以提升的。尽管他最初选择从事眼科（据报道是因为接触血液的情况较少），但阿萨德对大约 500,000人的死亡和失踪负有责任。

关键词：心理传记学，领导力分析，医生独裁者（doctator），独裁者，威权主义者，远距离评估

"Doctators" – Physicians Who Become Political Dictators

If you have a doctor who cuts the head because of gangrene to save a patient, you don't say he's a brutal doctor, he's doing his job in order to save the rest of the body. So when you protect your country from the terrorists and you kill terrorists, you are not brutal, you are a patriot. That's how you look at yourself, and that's how the people want to look at you. (Bashar al-Assad, President of Syria in Neely, 2016).

Over the years, some doctors have found their way into politics, serving in roles like a head of state or cabinet members. Several traits make doctors well suited for political office, including time management skills, making quick decisions under stressful situations, and working as a team (Stanley 2020). However, upon entering politics, doctors are looked upon differently compared to career politicians, lawyers, or any other professions. Arguably they are looked upon as father figures, persons to make us feel safe and secure (Perper and Cina 2010). Additionally, society perceives that through their wisdom and healing hands, they will know what is best for the country (Perper and Cina 2010). However, even some doctors have a dark side despite these high standards. The unique phenomenon of a physician turned dictator, coined "doctator," in the domains of political science and psychology was first identified and discussed by the journalist Robert Montefiore (1997), who defined the concept as "the process by which a medical doctor, devoted to sacrificing himself to save lives, becomes a dictator, devoted to sacrificing lives to save himself" (17). While several physicians have served honorably at the level of President or Prime Minister, including Sun Yat-Sen (First Provisional President, Republic of China), Ram Baran Yadav (President of Nepal), Gro Harlem Brundtland (Prime Minister of Norway), Michelle Bachelet (President of Chile), and Juscelino Kubitscheck (President of Brazil), elevated paternalistic roles in society have at times transferred to malignant national leadership roles by physician politicians (Lass et al., 2012).

Examples of previous physician dictators include Hastings Banda, Prime Minister, then President of Malawi. Banda, trained in medicine in the United States and Scotland, was reputed to have fed his imprisoned political enemies to crocodiles (Forster 2001). Francois "Papa Doc" Duvalier, President-for-Life of Haiti, transformed from an American-trained physician specializing in infectious and tropical diseases to a political dictator. Utilizing a cult of deified personality, he devised a combination of Voodoo, Christianity, torture, and extreme punishment during his 14-year reign of terror (Drumhiller and Skvorc 2018). Radovan Karadzic, President of the Srpska Republika, a Bosnia psychiatrist trained in Denmark and the United States, has been referred to as the "Butcher of Bosnia" for his actions resulting in a conviction of genocide and crimes against humanity by the International Criminal Tribunal for the former Yugoslavia.

Previous research focusing on doctators has provided additional insight into whether leaders are born or made and how physicians can come to commit and order extraordinary acts of violence against others (Drumhiller and Skvorc 2018). Drawing upon psychobiography as an at-a-distance assessment technique (Schultz 2005), this case study assesses how Bashar al-Assad has transitioned from a pathway of professional healing as an ophthalmologist into a doctator whose regime is responsible for the deaths and disappearances of more than 500,000 people (Ebrahimi 2021). In the early days of his rise to power, there were hopes that Bashar al-Assad would be very different from his father, bringing positive reforms to Syria and its people. This initial optimism can be attributed, in part, to the social bias and elevation afforded to him as a result of his medical training and practice as an Ophthalmologist (Perper and Cina 2010). Further fueling this hope, his inaugural speech contained assurances of "creative thinking, constructive criticism, transparency, and democracy" (Post 2015, 209). As a demonstration of this for a brief period, Assad allowed the organization of opposition parties, provided more freedoms to the press, and released many political prisoners as a show of good faith (de Halvetang 2017). Additionally, "liberal intellectuals founded discussion salons across the Syrian capital and put together political pamphlets and petitions for reform" (de Halvetang 2017, 4). However, these reforms were brief as Assad would later backtrack on his position and consolidate his power to maintain his position as President of Syria.

Psychobiography as a Means to Assess Doctor-Dictators

This study uses psychobiography as an at-a-distance assessment technique to assess Bashar al-Assad's transformation from ophthalmologist to a political dictator. As the term implies, using psychobiography to assess a political leader involves applying psychological concepts to a personal biographical history to better understand their leadership behavior and ultimately "make predictions about the individual's motivations'" (Cara 2007, 115; see also Lasswell 1930; Post 2003; Schultz 2005). Psychobiography has been used to assess a wide range of political leaders such as Woodrow Wilson (George and George 1998; McDermott 2008), Condolezza Rice (Fitch and Marshall 2008), Hitler (Victor 1998), and François Duvalier (Drumhiller and Skvorc 2018) to name a few. Furthermore, case studies of dictators have been carried out to assess their leadership transformation (Victor 1998); personalities (Coolidge and Segal 2007); psychopathology (Coolidge, Davis, and Segal 2007); as well as their overall development and pathology (Hyland, Boduszek, and Kiełkiewicz, 2011).

By analyzing written material focused on Assad's upbringing, socio-political development, and rise to power, we can assess his political psychological make-up by way of the personality traits that are revealed within the available data (Birt 1993; Coolidge, Davis, and Segal 2007; Coolidge and Segal 2007; 2009). This assessment considers Syria's development under Assad's father and how its organizational instabilities may have contributed to Assad's development. We review Assad's early years in power, hints of reform, and later his consolidation of power and human rights abuses. As a doctor-dictator, we pay particular attention to the development and cultivation of his medical interests, especially related to his later uses of this skillset to gain favor in the global community. This case study reveals Assad's use of Western-based professional training and sophistication to advance and justify the brutal treatment of any person or group perceived as disloyal. As a means of gaining additional insight into the doctor-dictator framework, this case study helps to identify the convergence of psychological traits and historical circumstances resulting in human rights violations on an epic scale.

Personality and Environmental Constructs

Dark Traits

Psychiatrists and psychologists have an established history of studying dictators both from a distance like that of Saddam Hussein (Post 2003; Coolidge and Segal 2007), Kim Jung Il (Coolidge and Segal 2009), or Stalin (Birt 1993; Stal 2003), and also in-person, such as Carl Jung's description of meeting Adolf Hitler and Benito Mussolini (Goldman 2011). In 2009, Coolidge and Segal identified six personality disorders common to dictators; these include the sadistic, antisocial, paranoid, narcissistic, schizoid, and schizotypal personality disorders. Over time, these disorders have come to be referred to as so-called "Dark Traits," which are reflected in dictators' cruel and antisocial behaviors (Haycock 2019). These traits include egoism, Machiavellianism, moral disengagement, narcissism, psychological entitlement, psychopathy, sadism, self-interest, and spitefulness. Examples include disregarding valid criticism, exploitation of others to facilitate an identified goal, holding oneself above moral and ethical conventions, extreme self-centeredness, extreme desire for reward without effort, lack of empathy and remorse for manipulative and dishonest behaviors, taking pleasure in abusing others (psychologically or physically), seeking rewards without concern of impact on others, and deliberate injuries to others.

Dark Traits have been incorporated in research to form a "common core of dark traits," known as a D-factor (Moshagen, Hilbig, and Zettler, 2018). In particular, "persons with a high D-factor score are capable of doing whatever is necessary to get what they want no matter who is inconvenienced or harmed .... They may act as if they want to help others, but their only goal ultimately is to help themselves" (Haycock 2019, 66). Associated behaviors include an exaggerated sense of self-importance, craving for excessive admiration, a self-entitled view of the world, willingness to exploit others, an absence of empathy, envy toward others perceived as successful, and an exaggerated self-report of accomplishments and achievements (Haycock 2019).

Dictators "see themselves as 'very special' people, deserving of admiration and, consequently, have difficulty empathizing with the feelings and needs of others .... Not only do dictators commonly show a 'pervasive pattern of grandiosity,' they also tend to behave with a vindictiveness often observed in narcissistic personality disorder" (Norrholm and Hunley 2017, 1). Politics are viewed as a means to personal gain, regardless of utilized methods or consequences to all others. Accompanying these perspectives is a willingness to believe conspiracy theories focused against them and a willingness to order extreme brutality (including death) to anyone challenging their authority. Subordinates are provided incentives, including status and power, to reinforce their loyalty, with disfavor resulting in extreme punishment, including incarceration or execution.

Additionally, autocratic leaders have difficulty experiencing empathy, love, and guilt (Burkle 2019). Instead, they are "distinguished by sociopathic and narcissistic behaviors that self-serve to cover their constant fear of insecurity and the insatiable need for power" (Burkle 2019, 1). In this regard, fundamental components within a democratic state, such as human rights and basic human security needs, are disregarded (Burkle 2019). Burkle further observed that dictators reflect a narcissism that exists concurrently with antisocial personality disorder, with a lack of conscience and remorse for their despotic behaviors and wrongdoings.

> Failing to govern fairly, substituting deviousness and deception for the lack of competence, obsessing over power, perpetuating fraud, and gaming the system become their own 'theology' .... Despite initial promises to the contrary, authoritarian leadership is too often based on manufactured fears, mistrust, and paranoia leading to incremental assaults on the population's human rights (Burkle 2019, 6).

Despite the work carried out linking personality disorders and dark traits to dictators, it is crucial to keep in mind that there are numerous examples of individuals with these six concurrent personality disorders who do not become dictators or terrorists, and some dictators likely have different diagnoses of personality and psychiatric disorders (Goldman 2011). Also, along these lines, having some of these darker traits is not enough to assume and sustain dictatorial power (Haycock 2019). As a result, it is also essential to consider the environmental conditions in the area when assessing leadership development and behavior.

Environmental Conditions

Instilling fear and terror in their opponents, lacking any remorse for the suffering of anyone challenging their rule, and appointing subordinates who can manage the logistics and details of administration—with a full understanding of their loss of societal privileges if there is a regime change—are tactics consistently used and reinforced by dictators. On their own, and sometimes even combined, these tactics are not enough to secure, let alone keep an individual in political power. This is likewise the case when it comes to their personality and personality traits. Moghaddam (2013) made the critical observation that comprehending the psychology of the dictator was less important than considering the environmental conditions that facilitate the rise of a dictator. Reinforcing this analysis, Haycock (2019) wrote, "it takes a broken nation to raise a dictator" (240). Facilitating events, including economic devastation, political revolution, and polarizing political divisions, can create situations ripe for the rise of individuals with the requisite personality traits and ambitions. Intense uncertainty, insecurity, and destabilization can lead to a willingness to believe messages of scapegoating, resentment, suspicion, and fear, where an "other" is blamed for the nation's problems (Haycock 2019; Roulet and Pichler 2020).

Freud's theory of displacement of aggression has been used to misdirect hostility to an innocent person or group to avoid appropriate channeling of responsibility (Haycock 2019). Likewise, discrediting the credibility of others and the use of a scapegoat also serve as additional blame avoidance strategies leaders may use to direct hostility towards another target (Preston 2011). Leading followers to fear an opposing group propels the dictator who promises protection against the identified group to power. Assertions of the inferiority of others have been linked to higher self-esteem and positive distinctiveness (Haycock 2019). The certainty of a superior national infrastructure that denies privileges and rights to some individuals or groups reinforces the sense of dominance, entitlement, and superiority (Haycock 2019). Trusting a dictator to properly interpret news events that potentially disagree with the dominance or superiority of the favored group reinforces the belief that only the dictator can preserve the status quo. Severe punishment for opposition sustains fear and obedience to the dictator (Haycock 2019).

> When a ruthless desire for power is combined with paranoid personality traits that help ensure survival, and a Machiavellian determination to do whatever is necessary to achieve power, you have a potential tyrant. Add in an ability to attract and hold the devotion of a set of followers, a messiah complex, or narcissistic belief of one's own greatness, and organizational skills, and you have a serious threat given the right circumstances for the tyrant. And you have the wrong circumstances for the rest of us (Haycock 2019, 239).

Maintaining a cult of personality requires the modern dictator to create an illusion of popular support by manipulating the electoral system, relying on secret police forces threatening loss of life or torture to assure maintenance of power (Gallo 2020).

In 1964, Erich Fromm coined the term "malignant narcissism," describing it as a severe mental illness and the personification of evil (Zeiders and Devlin 2020, 3). The concept of evil was used in the context of a socially cancerous generation of hate, culminating in terror, humiliation, and death. Related criminal features include criminal behavior, lying, betrayal, humiliation, brutality, and violation of civil rights. Paranoia leads to enemy identification and creates a bond of tribal exclusion. Sadism creates a deserving victim, for whom punishment is seen as a victory. Public calls for accountability evoke rage, fear, and loathing. Claims of infallible superiority fused with god-like powers of strength and authority are recurrent themes in speeches and interactions with followers.

Discussion

National and Political History

The region known historically as Syria is currently composed of the nations of Syria, Lebanon, Jordan, and Israel. In the past 4,000 years, foreign influence has been exerted by the Phoenicians, Aramaeans, Greeks, Romans, Muslim Arabs, European Christian Crusaders, Ottoman Turks, and the French. In addition, Alexander the Great and successive empires have exerted various cultural influences, including Christianity and Islam (Country Profile: Syria 2005). "Syria witnessed 15 successful *coup d'etats* between 1949-1970, external wars with Israel (1948, 1967, and 1973), vicious Pan-Arab competition with regional states, and a near civil war between 1976-1984" (Stacher 2011, 197). In modern times, Syria briefly merged with Egypt into the United Arab Republic (1958-1961). Syria subsequently broke away from Egypt, and Hafez Al-Assad was elected President of Syria in 1970. His 30-year Presidential term reflected an intolerance for dissent, enactment of a police state, and suppression of opposing political parties and leaders (Country Profile: Syria 2005). Hafez ultimately transformed Syria from a "coup-ridden 'semi-state into a veritable model of authoritarian stability'" (Stacher 2011, 197). In 1982, a political insurrection in Hama, the Syrian military, led by Assad's brother Rifaat and the Defence Brigades, regained control by destroying half the city and killing 10,000 residents (BBC 2012). During his rule, Hafez al-Assad established a highly personalized political regime which left the country with policies out of alignment with leadership practices (Stacher 2011; Owen 2014). For example, "Article 85 of the 1973 constitution provided that it was the first vice president, [then] Abdul-Halim Khaddam, who was to succeed the president on his death" (Owen 2014, 85). It is under this environmental backdrop where Bashar al-Assad would eventually gain power and control. However, ascension to the Syrian presidency was not his first career choice.

Family Background, Medical Training, Succession to Presidency

Bashar Hafez al-Assad was born on September 11, 1965, in Damascus (Al Jazeera 2018). His father was, at the time, Commander of the Syrian Air Force, and his mother belonged to a prominent family with extensive involvement in business and commerce. Bashar was the third child and second son. Described as a shy adolescent, he graduated from a private Arab-French high school before beginning his studies at the Damascus Medical University, focusing on the study of ophthalmology. Interestingly, his father, Hafez al-Assad, had initially wanted to study medicine; lacking the necessary monetary resources, he joined the military instead (Post and Pertsis 2011).

After an initial assignment to a local military hospital, Bashar moved to London to pursue advanced training at the Western Eye Hospital. In London, he met and married a British-Syrian investment banker working at JP Morgan, and the couple lived a quiet life. Assad has commonly been described as a with words and phrases such as being "gentle," "Westernized," and as being "computer nerd" or "geeky I.T. guy" (Hemmer 2003; Burke 2015). In some respects, he may have been unassuming as "he was the eye doctor, [who] spent time in London, [and] had the British-Syrian wife .... He was by and large completely separate to the politics of the country" (Burke 2015, 4). A friend described him as shy, covering his mouth when speaking and not making eye contact (de Haldevang 2017). He reportedly avoided large gatherings of people and would deliberately hunch over to minimize his tall height (de Haldevang 2017).

While Bashar was in England preparing for a medical career, his elder brother Bassel (three years older) was known as the successor to his father as President of Syria. "He was forceful, macho, an aficionado of fast cars who was popular with women. He stood in stark contrast to Bashar ... who grew up in Bassel's shadow, weak and in his own world, calm with a soft voice" (Post and Pertsis 2011, para. 4). Invoking family and national pride, President Hafez al-Assad referred to himself as "Abu Bassel" (father of Bassel), a label denoting the pride he took in his firstborn and chosen successor (Dwyer 2013).

Following the unexpected death of Bassel, Bashar was chosen by his father to be his new successor (Post 2015). Currently training in London to be an eye-surgeon, Bashar terminated his medical studies, returned to Syria to enroll in the Syrian Military Academy, and was quickly elevated to senior rank (Post 2015; Al Jazeera 2018). As President Hafez al-Assad became progressively burdened with severe health issues (heart disease and diabetes), Bashar's political profile in domestic and international politics and diplomacy became increasingly prominent. In order to ensure an effective turnover, Hafez al-Assad, retired or replaced several army offices and security chiefs that presented a possible threat to his son taking control of Syria (Rais 2004). Upon the death of President Hafez al-Assad, the Constitutional Presidential age requirement was reduced from 40 to 34, and Bashar al-Assad became President following an election in which he was the only candidate (Hemmer 2003; Owen 2014; Fares 2014).

Bashar al-Assad's early years, roughly between 2000 and 2001, were marked by moderate reform indicators and a move away from his father's grip on political organization. Key aspects of this included a brief ending to the Ba'ath monopoly over newspapers and political organizations. He also downplayed the cult of personality built-up around his father, Basil, and then later himself by discouraging the display of their pictures in public spaces (Hemmer 2003). Early on, Bashar also had a keen desire to hold a presidential election after his original seven-year term was up instead of another referendum (Hemmer 2003). Early characterizations of Bashar were largely positive, with him being presented as someone committed to the modernization of Syria, selflessly devoted "to his people," and also someone who seemingly had "wide support at home" (Heydemann 2007, 169). However, by 2001 Bashar can be seen to roll back his original attempts at openness, and by 2003, Bashar "embraced his role as America's adversary, shed his support for domestic political reform, and adopted a harsh, bellicose tone toward both the United States and Israel" (Heydemann 2007, 170). The consolidation of power would continue in 2005 when Assad forced the resignation and later exile of the Vice President, and in 2006 he would reshuffle his cabinet (Lesch 2010). Despite Assad's promises for economic and political reforms, in the end, instituting these with any real meaning for the people of Syria would result in a loss of power for the Assad regime and family as a whole (Hemmer 2003; Rais 2004).

Initial Indicators of Reform and Perception of Westernization

> *I am president, (but) I don't own the country, so (the military) are not my forces.* (Bashar al-Assad, in Post and Pertsis 2011).

At the beginning of Bashar Al-Assad's Presidency, indicators of political reform and freedom of political expression emerged as indicators of changes to his father's autocratic regime (Fisher 2012). Between September 2000 and February 2001, Assad allowed for moderate freedoms and political activity. Given his interests in technology, observers expected Assad's reign to be more open and transparent in the information age (Owen 2014). As an example of moderate freedoms, the "Manifesto of 99" was signed by Syrian intellectuals which "called for an end to the state of emergency in place since 1963" (Rais 2004, 156). In addition to releasing several political prisoners, he also provided "a substantial rise [sic] in the wages and salaries of public-sector workers, and [promised] to reactivate the role of the National Front, a defunct seven-party coalition led by the Ba'th" (Owen 2014, 86). Control over Syrian newspapers, previously monopolized by members of his political party, was relaxed, and other political parties were permitted to publish their newspapers. Assad even extended these journalistic reforms by allowing the publication of a satirical newspaper (Hemmer 2003). When proposals of political reforms were set forth and civil society forums emerged to discuss political change, there were no government reprisals (Hemmer 2003).

"Initially, Syrians and Syria-watchers hoped that Bashar would be an open-minded, liberal, and reforming leader. But these hopes rested on a fragile foundation" (Post and Pertsis 2011, para. 5). This optimism was based upon Assad's study of ophthalmology in London and his marriage to British-born Asma Assad, a graduate of Kings College and employee of the investment banking firm of JP Morgan, whose parents had immigrated from Syria. Subsequent political analysts found these assumptions to be superficial.

> Bashar was 27 when he lived in London, a fully formed adult and had spent his life absorbing his father's political ideas and observing his leadership style, in particular how to deal with conflict .... He only spent about 18 months in London, so his actual exposure to "Western" ways of life was likely quite limited ... and not a guarantee that (he would) adopt and internalize its values and ideals (Post and Pertsis 2011, para. 6).

The apprehension of growing political opposition stemming from the Damascus Spring resulted in a hard stop on reforms within the first year of the new Presidency (Owen 2014). An official announcement was made requirement advance government approval of all political meetings, new restrictions impacting the press and promulgation of news, and the arrest of emerging political leaders, including two members of Parliament, subsequently convicted of "aiming to change the Constitution by illegal means" (Hemmer 2003, 227). To prove himself as the strong successor to his father's autocratic Presidency, Bashar tightened his grip on the Presidency, and squashed any challenge to his authority (Post and Pertsis 2011; Miller and Sly 2021). In a drive against 'corruption' Bashar al-Assad gradually stripped his political opponents of their responsibilities, and others "such as senior generals and security chiefs whose loyalty to Bashar's succession was deemed suspect, were retired at the regulation age" (Owen 2014, 84).

The devolution of political rights has been linked to catastrophic declines in the Syrian economy. Arrests of business executives and seizure of corporate assets have been termed a "mafia-style money grab" (Miller and Sly 2021, para. 9). The current regime faces economic bankruptcy due to domestic civil war, debts to Iran and Russia, and the impact of continuing economic sanctions from the U.S. and other democratic nations. Syrian currency has been devalued 85% in the past decade. The current bleak economy has reduced the capacity of Syrian political rivals to challenge Assad's control. Assad has characterized government seizures of business assets as an anti-corruption measure, stating that "ending corruption is an economic, social and patriotic necessity" (Miller and Sly 2021).

It has recently been reported that an illegal drug industry has emerged in Syria, that is overseen by the Fourth Armored Division of the Syrian Army, under Bashar's younger brother Maher al-Assad (Hubbard and Saad 2021). Narcotics have been identified as the country's top export, with over 250 million pills seized worldwide in 2021. "The idea of going to the Syrian government to ask about cooperation is just absurd .... It is literally the Syrian government that is exporting the drugs .... They are the drug cartel" (Hubbard and Saad 2021). One business leader, reported to facilitate the movement of illegal narcotics, is a public supporter of the Assad regime and was recently awarded the Order of Merit by Assad, "in recognition of his prominent services in economics and financial management during a time of war" (Hubbard and Saad 2021, para. 35). The U.S. has imposed economic sanctions against Bashar and Maher al-Assad, and other identified Syrian participants in the trafficking of illegal drugs (Hubbard and Saad 2021).

Human Rights Violations and War Crimes

> *When asked, 'As you know, doctors take an oath never to do harm to anyone .... Does a doctor give them up when he takes office?' President Assad responded: 'First of all, a doctor takes the right decision to protect a patient. You can't say they do not do harm physically. Sometimes they have to extract the bad member that could kill the patient Extract eyes, could extract a leg, and so on, but you don't say he's a bad doctor. It's still a humanitarian job, whatever they do. The same for a politician but on a larger scale. Whether your decision helps the life of the Syrians or not in such a situation* (Resnick 2013, para 6-7).

Quality of life for the majority of Syrian citizens has starkly declined during the Assad regime. An estimated 700,000 Syrians have been killed or have disappeared during this time. The United Nations (2021b) has described Syria as the "world's largest refugee crisis," with more than six million Syrians escaping to foreign countries (UNHCR 2021). In addition, ninety percent of Syrians live in poverty, and 50% of Syrian children are malnourished. Despite these facts, Assad has steadfastly denied wrongdoing, characterizing his right to discipline his people "as a father would do to his son. 'The father is allowed to do whatever when the sons make mistakes ... this is a social contract between the Syrians and elected officials'" (de Haldevang 2017).

There have been a reported 336 chemical weapons attacks throughout the Syrian civil war, with 98% of these attributed to the Assad regime (Schneider and Lutkefend 2019). These attacks have been described as targeting opposing civilians (Schneider and Lutkefend 2019). Physicians for Human Rights has verified reports of 509 attacks made by the Syrian government or their allies on 348 medical facilities during 2011-2020. During this time, approximately 263 physicians and an additional 667 health care providers were killed, with an additional 143 medical professionals kidnapped, detained, or subsequently killed (Physicians for Human Rights 2019). The World Health Organization reports that Syria accounts for 70% of all healthcare facilities attacks documented worldwide (WHO 2018). In addition, an estimated 70% of the health workforce has left Syria (IRC 2021).

Over 80% of the Syrian population earns less than $1 daily. From 2014 to 2019, it has been reported that 40% of public schools have been destroyed and that over 2 million Syrian children do not have access to schools (Ghitman 2018).

During the past decade, at least 300 journalists have been arrested, and over 100 are missing (Reporters Without Borders 2021). The Syrian Network for Human Rights claims Assad is responsible for 14,338 deaths due to torture (including 173 children) in the past decade (Syrian Network for Human Rights 2021). After the defection of a member of the Syrian military police, 55,000 photographs were released, documenting the torture of an estimated 11,000 dead individuals held in Syrian detention centers. "The photographs were taken apparently as part of a bureaucratic effort by the Syrian security apparatus to maintain a photographic record of the thousands who have died in detention since 2011 as well as members of security forces who died in attacks by armed opposition groups" (Safi 2019). Assad responded to the release of the photographs and subsequent multi-national inquiry team with an assertion the materials were "allegations without evidence." However, a response by a former conscript at one of the military hospitals claimed otherwise. "I know this place from the photographs, stone by stone, brick by brick. I lived there 24 hours a day. I had to carry (the bodies) myself" (Motaparthy and Houry 2015).

Russian support of Syria in the United Nations has prevented allegations of Syrian war crimes to the International Criminal Court (Phillips 2021). In addition, international sanctions issued by the U.S., EU, Canada, Australia, and Switzerland have significantly impacted Syria's economy, but do not yet appear successful in deterring crimes against humanity (United Nations 2021a).

Malignant Narcissism and Dark Personality Traits

The preceding sections reflect Bashar al-Assad's presidency as a dictatorship maintaining brutal political power and control by imposing well-documented state-sanctioned human rights violations and engagement in war crimes (Moghaddam 2013). Central to Assad's dictatorship perpetuates the psychological constructs of malignant narcissism and dark personality traits. These constructs include the absence of "constraint of conscience, paranoid orientation, preoccupation by one's brilliance, and such extreme grandiosity that there is no capacity to empathize with others" (Post 2015, 190-191). Further, "dark personality traits" have been applied to describe dictatorships and include "messianic ambition for unlimited power, absence of conscience, unconstrained aggression, extreme self-interest, consistent use of deception, moral disengagement, psychological entitlement, sadism, and spitefulness" (Haycock 2019, 64-66).

Zeiders and Devlin (2020) describe malignant narcissism in political leaders as including identification with an ultimate power value, reflecting Assad's use of the analogy of himself as the physician who removes a patient's gangrenous body part, as reflected in his suppression of his political opponents (Resnick 2013). Machiavellianism, including false imprisonment and torture to suppress dissension, is an additional component of malignant narcissism, as is the creation of group scapegoats to enhance the cohesion of the predominant group. Associated perceptions of a malignant narcissistic dictator include, "I alone can solve this problem," "Diseases must be eradicated," and "They plot against us" (Zeiders and Devin 2020, 27-30).

In addition to these constructs, Moghaddam (2013) adds an interesting example of how the President and First Lady of Syria attempted to manipulate Western perception, specifically younger individuals, through an article in a fashion magazine published in 2011 (Buck 2011). The article described Mrs. Assad as a "rose in the desert" (529), and as "glamourous, and chic" (529), with a characterization of Syria as "the safest country in the Middle East" (Buck 2011, 529). The article claimed that the Assads saw no need for bodyguards and quoted Bashar as claiming their Hollywood guests, Brad Pitt and Angelina Jolie, wanted to "send his security guards here to get some training" (Buck 2011, 532). Moghaddam (2013) pointed out that Assad, whose dictatorship "relies on guns, tanks, tear gas and torture; he enforces obedience and conformity through direct, brutal tactics" (125), used the fashion article to deliberately intertwine his regime with the subtle dictates imposed by a media publication designed for young women aspiring to conform with popular public figures and fashion. Moghaddam (2013) noted that widespread objection to the article led to a rapid public relations decision to remove the article from website access.

Beginnings of Criminal Accountability

In a landmark case alleging crimes against humanity perpetrated by former members of Syria's General Intelligence Directorate, a German court has convicted and sentenced Eyad al-Gharab, a former low-ranking member of the Syrian General Intelligence Directorate, to a 4.5-year prison term. Additionally, Anwar Raslan, a former Colonel in the Syrian intelligence service, was also found guilty of human rights abuses and was sentenced to life in prison (Hubbard 2022). Al-Gharib was charged with bringing at least 30 protestors to a Damascus prison for torture in 2011. Raslan was convicted for his involvement in the torture of at least 4,000 Syrian prisoners and is charged with 58 counts of murder, with additional counts of rape and sexual assault.

The arrest warrants, charges, and trial were based upon a series of criminal complaints regarding torture in Syria in Germany, Austria, Sweden, and Norway. Germany is currently home to more than 800,000 Syrian refugees. Trial witness testimony described the killing and torture of prisoners on "almost an industrial scale." Witnesses described being physically beaten in prison, with torture techniques that included rape, hanging from the ceiling for hours, having their fingernails ripped out, the administration of electric shock, cigarette burns, and blows to the genitals (Amos 2021). In addition, Germany reportedly investigates dozens of former Syrian officials accused of atrocities (BBC News 2021).

Both defendants initially received asylum in Germany after fleeing Syria (Amos 2021). Al-Gharab's defense cited his fear of retribution and the killing of his family if he refused his orders and that he had assisted in the prosecution of his co-defendant, former Colonel Anwar Raslan (Amos 2021). Both verdicts have been characterized as a significant test-case to establish a record of evidence regarding the crimes against humanity committed by the regime of President al-Assad. These cases help send the message that people can be held accountable for their actions, further conveying a message of future accountability for Assad when he is no longer President, and for those continuing to serve the Assad regime. Russia and China have, so far, thwarted actions by other nations to establish an international tribunal for Syrian war crimes (Amos 2021).

Conclusion

Bashar al-Assad, MD, and his father, Syrian President Hafez al-Assad, share remarkable parallel career outcomes. Hafez's original career goal was to become a physician; lacking the necessary funds to attend medical school, his alternate choice was a military career that ultimately catapulted him to the country's Presidency (Seale 2000). His autocratic regime was characterized by brutal suppression of political opposition and human rights violations (Post and Pertsis 2011).

Born into wealth and privilege, Bashar's desire to become a physician was easily assured, progressing to graduation from medical school and placement in a London ophthalmology medical residency (Burke 2015). However, after the death of his elder brother Bassel, the original heir apparent to succeed his father, Bashar set aside his medical studies, returned to Damascus, and enrolled into the Syrian Military Academy. Upon the death of Hafez from a heart attack, the Syrian Constitution was rapidly amended to lower the age requirement for office. As a result, Bashar was elected President of the Syrian Arab Republic, winning 97% of the vote (Al Jazeera 2018).

At the time of his succession/election, it was believed that Bashar's training as a physician could potentially result in a more progressive presidential administration, something depicted in the statement, "we've lost a dictator, and gained an ophthalmologist" (*The Economist* 2000). Soon thereafter, Bashar removed any ambiguity regarding his intent to closely emulate his father in doing whatever necessary to cement his dictatorial authority (Fisher 2012).

Bashar's invocation of a medical analogy to justify his brutality reflects his abiding self-perception as an omnipotent physician/president who knows best, regardless of outcome measures. Unlike a sick patient who hopes to restore their health after surgery, quality of life measures for the nation of Syria has undergone a catastrophic decline during the Bashar administration. For the Bashar Al-Assad presidency, the historic physician admonition of Hippocrates to "first, do no harm" has been systematically violated.

References

Al Jazeera. 2018. "Profile: Bashar al-Assad." April 17, 2018. https://www.aljazeera.com/news/2018/4/17/profile-bashar-al-assad.

Amos, Deborah. 2021. "Landmark Verdict in Germany Sentences Syrian for Aiding Crimes Against Humanity." *NPR*, February 24, 2015. https://www.npr.org/2021/02/24/970663111/landmark-verdict-in-germany-sentences-syrian-official-for-crime-against-humanity.

BBC. 2012. "Profile: Syrian City of Hama." *BBC News*, April 27, 2012. http://www.bbc.com/news/world-middle-east-17868325.

BBC. 2021. "Syria Torture: German Court Convicts Ex-Intelligence Officer." *BBC News*, February 24, 2021. https://www.bbc.com/news/world-europe-56160486.

Birt, Raymond. 1993. "Personality and Foreign Policy: The Case of Stalin." *Political Psychology* 14, no. 4: 607–25.

Buck, Joan Juliet. 2011. "A Rose in the Desert." *Vogue*, March 2011, 529-533, 571.

Burke, Sarah. 2015. "How Syria's 'Geeky' President Assad Went from Doctor to Dictator." *NBCNews.com,* October 30, 2015. https://www.nbcnews.com/storyline/syria-peace-talks-how-syrias-geeky-president-assad-went-doctor-dictator-n453871.

Burkle, Frederick M. 2019. "Character Disorders Among Autocratic World Leaders and the Impact on Health Security, Human Rights, and Humanitarian Care." *Prehospital and Disaster Medicine* 34, no. 1: 1–18.

Cara, Elizabeth. 2007. "Psychobiography: A Research Method in Search of a Home." *The British Journal of Occupational Therapy* 70, no. 3: 115–21.

Coolidge, Frederick L., and Daniel L. Segal. 2007. "Was Saddam Hussein Like Adolf Hitler? A Personality Disorder Investigation." *Military Psychology* 19, no. 4: 289–99.

Coolidge, Frederick L., and Daniel L. Segal. 2009. "Is Kim Jong-Il Like Saddam Hussein and Adolf Hitler? A Personality Disorder Evaluation." *Behavioral Sciences of Terrorism and Political Aggression* 1, no. 3: 195–202.

Coolidge, Frederick L., Felicia L. Davis, and Daniel L. Segal. 2007. "Understanding Madmen: A DSM-IV Assessment of Adolf Hitler." *Individual Differences Research* 5, no. 1: 30–43.

Country Profile: Syria. The Library of Congress. (2005, April). Accessed December 9, 2021. https://loc.gov/rr/frd/cs/profiles/Syria.pdf.

International Rescue Committee (IRC). 2021. "A Decade of Destruction: Attacks on Health Care in Syria." Accessed December 9, 2021. https://www.rescue.org/article/decade-destruction-attacks-health-care-syria.

de Haldevang, Max. 2017. "The Enigma of Assad: How a Painfully Shy Eye Doctor Turned into a Murderous Tyrant." *Quartz*, April 21, 2017. https://qz.com/959806/the-enigmatic-story-of-how-syrias-bashar-al-assad-turned-front-a-paintfully-shy-eye-doctor-into-a-murderous-tyrant/.

Drumhiller, Nicole K., and Casey Skvorc. 2018. "A Psychological and Political Analysis of a 20th Century 'Doctator': Dr. François Duvalier, President-for-Life of Haiti." *Global Security and Intelligence Studies* 3, no. 1: 9–32. https://viewer.joomag.com/global-security-and-intelligence-studies-volume-3-number-1-spring-summer-2018/0583559001528380568?short&.

Dwyer, Mimi. 2013. "Think Bashar al Assad is Brutal? Meet His Family." *The New Republic*, September 8, 2013. https://newrepublic.com/article/114630/bashar-al-assad-syria-family-guide.

Ebrahimi, Soraya. 2021. "A Decade of War in Syria Killed Over 388,000, Says British Human Rights Group." *The National,* March 15, 2021. https://www.thenationalnews.com/mena/a-decade-of-war-in-syria-killed-over-388-000-says-british-human-rights-group.

Fares, Qais. 2014. "The Syrian Constitution: Assad's Magic Wand." *Carnegie Middle East Center Diwan,* May 8, 2014. https://carnegie-mec.org/diwan/?lang=en.

Fisher, Marc, 2012. Syria's Assad has embraced pariah status. *The Washington Post,* June 16, 2012. https://www.washingtonpost.com/world/middle_east/syrias-assad-has-embraced-pariah-status/2012/06/16/gJQAsY9shV_story.html

Forster, Peter. 2001. "Law and Society Under a Democratic Dictatorship: Dr. Banda and Malawi." *Journal of Asian & African Studies* 36, no. 3: 275–293.

Fitch, Trey, and Jennifer Marshall. 2008. "A Comparative Psychobiography of Hillary Clinton and Condoleezza Rice." Conference Paper presented at the American Counseling Association's National Conference, Oahu, HI.

Gallo, Francesco. 2020. "Psychopathology of Dictators." *Il Sileno Edizioni,* July 24, 2020. https://www.ilsileno.it/rivistailsileno/2020/07/24/psychopathology-of-dictators.

George, Juliette, L, and Alexander George L. 1998. *Presidential Personality and Performance*. Westview Press.

Ghitman, Elise. 2018. 10 Facts About Poverty in Syria. *The Borgen Project,* May 4, 2018. https://borgenproject.org/poverty-in-syria-2/.

Goldman, Jason G. 2011. "The Psychology of Dictatorship: Kim Jong-Il." *Scientific American,* December 19, 2011. https://blogs.scientificamerican.com/thoughtful-animal/the-psychology-of-dictatorship-kim-jong-il?print=true.

Haycock, Dean A. 2019. *Tyrannical Minds: Psychological Profiling, Narcissism, and Dictatorship*. New York: Pegasus Books.

Hemmer, Christopher. 2003. "Syria Under Bashar al-Asad: Clinging to His Roots?" In *Know Thy Enemy: Profiles of Adversary Leaders and Their Strategic Cultures,* edited by Barry R. Schneider and Jerrold M. Post, 221–246. Maxwell, AL: Maxwell Air Force Base. http://hdl.handle.net/2027/mdp.39015053027457

Heydemann, Steven. 2007. "Review of the New Lion of Damascus: Bashar al-Asad and Modern Syria." *Political Science Quarterly* 122, no. 1: 168–170.

Hubbard, Ben. 2022. "Former Syrian Colonel Guilty in War Crimes Trial in Germany." *The New York Times,* January 13, 2022. https://www.nytimes.com/

live/2022/01/13/world/syria-war-crimes-germany-verdict.

Hubbard, Ben and Hwaida Saad. 2021. "On Syria's Ruins, a Drug Empire Flourishes." *The New York Times,* December 5, 2021. https://www.nytimes.com/2021/12/05/world/middleeast/syria-drugs-captagon-assad.html.

Hyland, Philip, Daniel Boduszek, and Krzysztof Kielkiewicz. 2011. "A Psycho-Historical Analysis of Adolph Hitler: The Role of Personality, Psychopathology, and Development." *Psychology & Society* 4, no. 2: 58–63.

Lass, Piotr, Adam Szarszewski, Aleksandra Gaworska-Krzemińska, and Jarosław Sławek. 2012. "Medical Doctors as the State Presidents and Prime Ministers: A Biographical Analysis." *Przeglad Lekarski* 69, no. 8: 642–46.

Lasswell, Harold D. 1930. *Psychopathology and Politics.* Chicago: University of Chicago Press.

Lesch, David W. 2010. "The Evolution of Bashar al-Asad." *Middle East Policy* 17, no. 2: 70–81.

McDermott, Rose. 2008. *Presidential Leadership, Illness, and Decision Making.* Cambridge, MA: Cambridge University Press.

Miller, Greg, and Liz Sly. 2021. "Assad's Tightening Grip." *The Washington Post,* December 4, 2021. http://www.washingtonpost.com/world/interactive/2021/assad-syria-business-government/.

Moghaddam, Fathali. M. 2013. *Psychology of Dictatorship.* Washington, D.C.: American Psychological Association.

Montefiore, Simon Sebag. 1997. "The Era of the Doctators." *The Spectator.* (April 12): 17–18.

Moshagen, Morten, Benjamin E. Hilbig, and Ingo Zettler. 2018. "The Dark Core of Personality." *Psychological Review* 125, no. 5: 656–88.

Motaparthy, Priyanka and Nadim Houry. 2015. "If the Dead Could Speak: Mass Deaths and Torture in Syria's Detention Facilities." *Human Rights Watch.* Accessed December 10, 2021. https://www.hrw.org/report/2015/12/16/if-dead-could-speak/mass-deaths-and-torture-syrias-detention-facilities#.

Neely, Bill. 2016. "Full Exclusive Interview with Syrian President Bashar Al-Assad." *NBC Nightly News,* August 19, 2016. https://www.nbcnews.com/news/world/

watch-full-exclusive-interview-syrian-president-bashar-al-assad-n634471.

Norrholm, Seth D. and Samuel Hurley. 2016. "The Psychology of Dictators: Power, Fear, and Anxiety." *Anxiety.org*, July 1, 2016. https://www.anxiety.org/psychology-of-dictators-power-fear-anxiety.

Owen, Roger. 2014. *The Rise and Fall of Arab Presidents for Life*. Cambridge: Harvard University Press.

Perper, Joshua A. and Stephen J. Cina. 2010. *When Doctors Kill: Who, Why and How*. New York: Copernicus Books.

Phillips, Roger Lu. 2021. "A Drop in the Ocean: A Preliminary Assessment of the Koblen Trial on Syrian Torture." *Just Security*, April 22, 2021. https://www.justsecurity.org75849/a-drop-in-the-ocean-a-preliminary-assessment-of-the-koblenz-trial-on-syrian-torture/.

Physicians for Human Rights. 2019. "Medical Personnel are Targeted in Syria." Accessed December 9, 2021. https://phr.org/our-work/resources/medical-personnel-are-targeted-in-syria.

Post, Jerrold M. 2003. *The Psychological Assessment of Political Leaders: With Profiles of Saddam Hussein and William Jefferson Clinton*. Ann Arbor, MI: University of Michigan Press.

Post, Jerrold M. 2004. *Leaders and Their Followers in a Dangerous World: The Psychology of Political Behavior*. Ithaca: Cornell University Press.

Post, Jerrold M. 2015. *Narcissism and Politics: Dreams of Glory*. New York: Cambridge University Press.

Post, Jerrold M. and Ruthie Pertsis. 2011. "Bashar al-Assad is Every Bit His Father's Son." *Foreign Policy*, December 21, 2011. https://foreignpolicy.com/2011/12/20/bashar-al-assad-is-every-bit-his-fathers-son/.

Preston, Thomas. 2011. *Pandora's Trap: Presidential Decision Making and Blame Avoidance in Vietnam and Iraq*. Lanham: Roman & Littlefield.

Rais, Faiza R. 2004. "Syria Under Bashar Al Assad: A Profile of Power." *Strategic Studies* 24, no. 3: 144–168.

Reporters Without Borders. 2021. "Toll of Ten Years of Civil War on Journalists in Syria." Accessed December 9, 2021. https://rsf.org/en/news/toll-ten-years-civ

il-war-journalists-syria#:~:text=According%20to%20data%20compiled%20%E2%80%93%20and,victims%20of%20education%20 since%202011.

Resnick, Brian. 2013. "How Bashar al-Assad Rationalizes 'Doing Harm.'" *The Atlantic,* September 18, 2013. https://www.theatlantic.com/politics/archive/2013/09/how-bashar-al-assad-rationalizes-doing-harm/454125/.

Roulet, Thomas J., and Rasmus Pichler. 2020. "Blame Game Theory: Scapegoating, Whistleblowing and Discursive Struggles Following Accusations of Organizational Misconduct." *Organizational Theory* 1, no. 4: 1–30.

Rudaw.net. 2021. "German Prosecutors Seek Life for Syrian in Torture Case." February 12, 2021. https://www.rudaw.net/english/world/021220212.

Safi, Michael. 2019. "Syrian Regime Inflicts 72 Forms of Torture on Prisoners, Report Finds." *The Guardian,* October 23, 2019. https://www.theguardian.com/world/2019/oct/23/syrian-regime-inflicts-72-forms-of-torture-on-prisoners-report-finds.

Schneider, Tobias and Theresa Lütkefend. 2019. "Nowhere to Hide: The Logic of Chemical Weapons Use in Syria." *Global Public Policy Institute.* Accessed December 9, 2021. https://www.gppi.net/2019/02/17/the-logic-of-chemical-weapons-use-in-syria.

Schultz, William Todd, ed. 2005. *Handbook of Psychobiography.* 1st ed. Oxford: Oxford University Press.

Seale, Patrick. 2000. "Hafez al-Assad: Feared and Respected Leader Who Raised Syria's Profile but was Ultimately Unable to Contain Israel." *The Guardian,* June 11, 2000. https://www.theguardian.com/news/2000/jun/12/guardianobituaries.israel.

Stacher, Joshua. 2011. "Reinterpreting Authoritarian Power: Syria's Hereditary Succession." *The Middle East Journal* 65, no. 2: 197–212.

Stal, Marina. 2013. "Psychopathology of Joseph Stalin." *Psychology* 4, no. 9: 1–4.

Stanley, Ben. 2020. "From Physician to Legislator: The Long History of Doctors in Politics." *The Rotation* (May 15). Accessed September 14, 2021. https://the-rotation.com/from-physician-to-legislator-the-long-history-of-doctors-in-politics.

Syrian Network for Human Rights. 2021. "The Tenth Annual Report on Torture in Syria on the International Day in Support of Victims of Torture." Accessed December 10, 2021. https://reliefweb.int/report/syrian-arab-republic/tenth-annual-report-torture-syria-international-day-support-victims.

The Economist. 2000. "Syria: Bashar's World." June 17, 2020, 355(8175), 24–26. https://www.economist.com/special/2000/06/15/bashars-world.

United Nations. 2021a. "Commission of Inquiry on the Syrian Arab Republic: Parties to the Conflict Continue to Perpetuate War Crimes and Crimes Against Humanity, Infringing on the Basic Human Rights of Syrians." OHCHR. Accessed December 10, 2021. https://www.ohchr.org/en/NewsEvents/Pages/DisplayNews.aspx?NewsID=27526&LangID=E.

United Nations, Secretary General. 2021b. "As Plight of Syrians Worsens, Hunger Reaches Record High, International Community Must Fully Commit to Ending Decade-Old War, Secretary-General Tells General Assembly." Secretary-General: Statements and Messages - Press Release. Accessed December 26, 2021. https://www.un.org/press/en/2021/sgsm20664.doc.htm.

Victor, George. 1998. *Hitler: The Pathology of Evil*. Dulles: Potomac Books.

World Health Organization. 2018. "Attacks on Health Care Are on the Rise Throughout Syria in the First Half of 2018, Says WHO." Accessed December 10, 2021. http://www.emro.who.int/syria/news/attacks-on-health-care-on-the-rise-throughout-syria-in-first-half-of-2018-says-who.html.

Zeiders, Charles, and Peter Devlin. 2020. *Malignant Narcissism and Power: A Psychodynamic Exploration of Madness and Leadership*. New York: Routledge.

Casey Skvorc, PhD, JD, is a medical psychologist and attorney, and serves as an Associate Professor in the doctoral programs of Strategic Intelligence and Global Security at the American Public University System. His primary focus of research includes the application of psychology, public health, and law to political psychology, law and ethics in the American Intelligence community, international public health and security, and medical intelligence. Highlights from his joint research with co-author Nicole K. Drumhiller include psychobiographies of physicians who became political dictators, including Francois Duvalier, President-for-Life of Haiti; Hastings Banda, Prime Minister and President of Malawi; and Radovan Karadzic, President of the Srpska Republika.

Nicole K. Drumhiller, PhD, CTM, is the Interim Dean of the School of Security of Global Studies at the American Public University System. She also serves as an Advisory Board member for the Operative Intelligence Research Center in Rome, Italy and is a co-Founder of the IntelHub, an online consortium for intelligence education and research. Her published works focus on the political behaviors of groups and individuals and covers the areas of political psychology, intelligence, threat management, religious cult behavior, and international relations. Her recent publications include *The Academic Practitioner-Divide in Intelligence Studies* (Roman & Littlefield, 2022, in press); "Advice, Decision Making, and Leadership in Security Crises" (Encyclopedia of Crisis Analysis, Oxford University Press; 2022); and "Religion and State Authority: Control of the Body" with Kate Brannum in the *Journal of Coercion, Abuse, and Manipulation* (2022, in press).

Global Security and Intelligence Studies • *Volume 7, Number 1* • *Summer 2022*

The Soft Path to U.S. Hegemony in the 21st Century: An American Brain Drain Policy against Strategic Competitors

Ryan Burke & Jahara Matisek

Disclosure: The views expressed are those of the authors and do not reflect the official position of the U.S. Air Force Academy, U.S. Naval War College, Department of the Air Force, Department of Defense, or any organization or institution with which the authors are affiliated.

Abstract

During the four years of the Trump administration, American strategy adopted a provocative *realpolitik* approach to American power. Trump's administration was so focused on U.S. nationalism and threats to American identity that domestic immigration policies became viewed as existential threats. With the introduction of great power competition discourse against China and Russia 2017, a renewed U.S. emphasis on conventional military power and traditional warfare ignores the reality of an increasingly globalized, interconnected world. China and Russia have grown their regional spheres of influence while making in-roads elsewhere with asymmetric tools of influence. While debates rage about how to confront China and Russia through projection of military and economic power, we ask: What can the U.S. do in the long-term to out-compete illiberal authoritarian states? Nullifying Chinese and Russian economic and military power means re-creating a grand strategy that leverages the unwelcoming internal illiberal politics in China and Russia. The U.S. must capitalize on the totalitarian world vision espoused by strategic competitors by emphasizing a liberal ideology to convince the best and brightest around the world to immigrate to America, thus contributing to innovation and overall American hard power. Choosing liberalism over realism will produce an attractive citizen recruitment proposition for the most educated and innovative citizens living in authoritarian regimes, with them leaving their home countries in favor of a United States that promotes free-markets and inclusivity. We contend that America needs a *Strategic Brain Drain* policy as a grand strategy pillar for 21st century strategic competition.

doi: 10.18278/gsis.7.1.5

Keywords: Brain Drain, Great Power Competition, China, Russia, Strategic Competition

El camino suave hacia la hegemonía estadounidense en el siglo XXI: Una política estadounidense de fuga de cerebros contra los competidores estratégicos

Resumen

Durante los cuatro años de la administración Trump, la estrategia estadounidense adoptó un enfoque provocativo de *realpolitik* para el poder estadounidense. La administración de Trump estaba tan concentrada en el nacionalismo estadounidense y las amenazas a la identidad estadounidense que las políticas de inmigración interna se consideraron amenazas existenciales. Con la introducción del discurso de competencia de las grandes potencias contra China y Rusia en 2017, un énfasis renovado de EE. UU. en el poder militar convencional y la guerra tradicional ignora la realidad de un mundo cada vez más globalizado e interconectado. China y Rusia han ampliado sus esferas regionales de influencia mientras se abren camino en otros lugares con herramientas asimétricas de influencia. Mientras se debate sobre cómo enfrentar a China y Rusia a través de la proyección del poder militar y económico, nos preguntamos: ¿Qué puede hacer EE. UU. a largo plazo para superar a los estados autoritarios no liberales? Anular el poder económico y militar de China y Rusia significa recrear una gran estrategia que aproveche la política interna antiliberal poco acogedora en China y Rusia. Estados Unidos debe sacar provecho de la visión totalitaria del mundo propugnada por los competidores estratégicos al enfatizar una ideología liberal para convencer a los mejores y más brillantes del mundo para que emigren a Estados Unidos, contribuyendo así a la innovación y al poder duro estadounidense en general. Elegir el liberalismo sobre el realismo producirá una atractiva propuesta de reclutamiento de ciudadanos para los ciudadanos más educados e innovadores que viven en regímenes autoritarios, dejando sus países de origen en favor de Estados Unidos que promueve el libre mercado y la inclusión. Sostenemos que Estados Unidos necesita una política de fuga de cerebros estratégica como un gran pilar estratégico para la competencia estratégica del siglo XXI.

Palabras clave: Fuga de cerebros, Gran competencia de poder, China, Rusia, Competencia estratégica

通往21世纪美国霸权的软路径：美国针对战略竞争对手的人才流失政策

摘要

特朗普执政的四年里，美国战略对美国实力采取了煽动性的现实政治（realpolitik）方法。特朗普政府如此专注于美国民族主义和对美国身份的威胁，以至于国内移民政策被视为生存威胁。随着2017年提出针对中国和俄罗斯的大国竞争话语，美国重新强调常规军事力量和传统战争，忽视了日益全球化、相互关联的世界现实。中国和俄罗斯扩大了地区影响力范围，同时利用不对称的势力手段影响其他地区。尽管关于"如何通过投射军事力量和经济力量以对抗中国和俄罗斯"的辩论迅速蔓延，我们研究的问题是：从长远来看，美国能做些什么来战胜非自由的威权主义国家？让中国和俄罗斯的经济实力和军事实力失效，这意味着重新制定一个利用中国和俄罗斯不受欢迎的内部非自由政治的大战略。美国必须利用战略竞争对手所拥护的极权主义世界观，通过强调自由主义意识形态，说服世界上最优秀、最聪明的人移民美国，从而为创新和美国整体硬实力作贡献。选择自由主义而不是现实主义，将为生活在威权制度中的、受教育程度最高且创新能力最强的公民提供有吸引力的公民招募提议，让其离开自己的祖国，转而支持推崇自由市场和包容性的美国。我们认为，美国需要战略人才流失（Strategic Brain Drain）政策作为21世纪战略竞争的大战略支柱。

关键词：人才流失，大国竞争，中国，俄罗斯，战略竞争

Already, with Russia's 2022 invasion of Ukraine, over 4 million educated Russians have fled the country, prompting the Biden administration to consider easing visa rules to capitalize on such massive brain drain of Russia.[1]

1 Zahra Tayeb, "The US plans to capitalize on a Russian 'brain drain' by easing visa requirements for workers with STEM skills, reports say," *Business Insider*, 30 April 2022, https://www.businessinsider.com/biden-plans-russian-brain-easing-visa-rules-stem-professionals-2022-4; Niko Vorobyov, "'Criminal adventure': Ukraine war fuels Russia's brain drain," *Al Jazeera*, 23 May 2022, https://www.aljazeera.com/news/2022/5/23/many-leave-russia-as-ukraine-war-drags-on.

Introduction

Throughout the rings of the Pentagon, strategists and policymakers are attempting to change American grand strategy for burgeoning 21st-century threats. Grand strategy is a reflection of national values, and includes policies and plans informing the application of instruments of power to advance national interests.[2] It establishes the roadmap a nation follows in its quest for international standing and power as demonstrated during the Cold War, where both political parties and foreign policy elites could mostly agree on American interests and threats to them.[3] However, when it comes to confronting adversaries, short-term politicking can undermine long-term planning for maintaining and/or growing American power.

After the Cold War, a bipartisan blend of hawks and doves saw a role in using military force to promote American values abroad. Such efforts were pursued by both political parties without much consideration to the externalities that this liberal use of American power would have in the long-term when intervening in the affairs of peripheral countries with little to no threat to U.S. national interests.[4] Though the Trump Administration continued these policies to an extent with continued military engagement in Iraq, Afghanistan, and Syria, the principal focus of its defense policies shifted toward Great Power Competition with the 2017 *National Security Strategy* (*NSS*) while viewing immigration as a vulnerability and threat to the homeland.[5] This new path altered the course of America's grand strategy, which generated debate among strategists and policymakers alike.[6]

In some ways, the focus of the 2017 *NSS* prevailed, evidenced in the Biden Administration's 2021 *Interim National Security Strategic Guidance* (*NSSG*) framing China and Russia as rivals in strategic competition with the U.S. At the same time, the *Interim NSSG* also rebuked the hostile anti-immigration undertones of the previous administration by advocating for "restoring our nation's historic strengths by ensuring our immigration policy incentivizes the world's best and brightest to study, work, and stay in America."[7] However,

2 John Lewis Gaddis, *On Grand Strategy* (New York: Penguin Press, 2018).

3 Hal Brands, "Rethinking America's grand strategy: Insights from the Cold War," *Parameters* 45, no. 4 (2016): 7-16.

4 Stephen M. Walt, "US grand strategy after the Cold War: Can realism explain it? Should realism guide it?" *International Relations* 32, no. 1 (2018): 3-22.

5 Donald J. Trump, *National Security Strategy of the United States of America* (Washington, D.C.: The White House, 2017).

6 Hal Brands, "The unexceptional superpower: American grand strategy in the age of trump," *Survival* 59, no. 6 (2017): 7-40; Barry R. Posen, "The Rise of Illiberal Hegemony: Trump's Surprising Grand Strategy," *Foreign Affairs* 97 (2018): 20-27.

7 Joseph R. Biden, Jr., *Interim National Security Strategic Guidance* (Washington, D.C.: The White House, 2021): 17.

the Trump administration damaged perceptions of U.S. 'soft power' through its comparably more restrictive immigration laws that included unprecedented resistance to U.S. immigration among even well-educated foreign professionals.[8] The Trump administration use of "America First," and policies that reflected such a narrow world view, further undermined the tremendous soft power the U.S. had accumulated over decades as a place for immigrants to move to and excel.

Besides the "Muslim Ban," the Trump Administration's crude demagogy approach to immigration, led primarily by Stephen Miller, resulted in dramatic changes to policies and laws.[9] For instance, by early 2021 legal immigration had been reduced by 49 percent, at least 30 percent of highly skilled foreign-born individuals had their H1-B visas rejected, whereas the typical rate used to be 6 percent, and there was a 143 percent increase in immigrants being denied naturalization despite their service in the US military.[10] Such shifts in domestic policies can have major ramifications as recent research shows that "despite immigrants only making up 16% of inventors, they are responsible for 30% of aggregate U.S. innovation since 1976, with their indirect spillover effects accounting for more than twice their direct productivity contribution."[11] Correlation is not causation, but consider China surpassing the U.S. in global patent applications for the first time ever in 2019, with China filing 58,990 applicants to beat out the U.S. rate of 57,840.[12] American power relies on innovation and advanced technologies.[13] The slipping of U.S. innovation dominance should be a source of strategic concern because it not only helps the U.S. retain an economic edge, but technological ad-

8 For instance, the Obama administration's *NSS* viewed immigration as contributing to the power of the U.S., while the Bush administration *NSS* barely mentioned immigration other than to deal with illegal immigration. Regardless, there was political consensus across the spectrum concerning the value of immigration to America. Desirée Colomé-Menéndez, Joachim A. Koops, and Daan Weggemans, "A country of immigrants no more? The securitization of immigration in the National Security Strategies of the United States of America," *Global Affairs* (2021): 1-26.

9 Julie Hirschfeld Davis and Michael D. Shear, *Border Wars: Inside Trump's Assault on Immigration* (New York: Simon & Schuster, 2020).

10 Stuart Anderson, "A Review of Trump Immigration Policy," *Forbes*, 26 August 2020, https://www.forbes.com/sites/stuartanderson/2020/08/26/fact-check-and-review-of-trump-immigration-policy/?sh=6c48b5e756c0.

11 Shai Bernstein, Rebecca Diamond, Timothy McQuade, and Beatriz Pousada, "The Contribution of High-Skilled Immigrants to Innovation in the United States," Working Paper, Stanford University (2019), 1.

12 Stephanie Nebehay, "In a first, China knocks U.S. from top spot in global patent race," *Reuters*, 7 April 2020, https://www.reuters.com/article/us-usa-china-patents/in-a-first-china-knocks-u-s-from-top-spot-in-global-patent-race-idUSKBN21P1P9.

13 Teryn Norris and Neil K. Shenai, "Dynamic balances: American power in the age of innovation," *SAIS Review of International Affairs* 30, no. 2 (2010): 149-164.

vancements support and facilitate weapon system developments.

The Trump Administration developed new rhetoric and threat framing through the *NSS*. The 2017 *NSS* realigned U.S. security priorities at home and abroad. It drove the Pentagon to think differently about new problem sets, and the policies, technologies, and force structures necessary to meet them. In particular, the Department of Defense's Joint Staff 12th annual Strategic Multilayer Assessment (SMA) project, *Future of Great Power Competition & Conflict*, attempted to further this strategic reorientation of American policy by surveying dozens of experts in the field to uncover strategic gaps and efficiencies.[14] One such question from the SMA explicates the precarious situation U.S. policymakers think they are in: *What are the long-term implications for the U.S. of adopting an objective of strategic parity with China and Russia rather than military dominance?*

While this question seems direct and unambiguous, the question by its very existence exposes a great flaw in reasoning about how the U.S. should handle China, a rising near-peer, and Russia, a declining revisionist state with a plummeting birth rate and life expectancy that under President Putin, *thinks* its 6,000 nuclear weapons bestow peer status with the U.S.[15] Instead, we argue that the SMA question should be revised, asking: *What can the U.S. do in the long-term so that it benefits from China and Russia being illiberal authoritarian states?*

All too often, American grand strategy looks outwards to existential threats and other U.S. national interests without acknowledging the domestic factors that have contributed to economic and military power over two centuries. Societal elements are sometimes overlooked when it comes to considering what allows a country from mobilizing all its power and resources. While many scholars traditionally rely on the size of the population base as one of the primary metrics, alongside urban population, iron and steel production, energy consumption, military expenditure and number of personnel in the military, to measure national power, such calculations miss data points such as population trends, birth rates, educational attainment, and immigration.[16]

As 2018 data indicated the U.S. experienced its lowest birth rate since 1982,[17] it is even more troubling that immigration trends to the U.S. in 2019

14 An archive of SMA projects dating back to 2007 is maintained here: https://nsiteam.com/sma-publications/.

15 Angela E. Stent, *Putin's World: Russia against the West and with the Rest* (New York: Hachette, 2019).

16 David J. Singer, Stuart Bremer, and John Stuckey, "Capability distribution, uncertainty, and major power war, 1820-1965," in: *Peace, war, and numbers*, edited by Bruce Russett (Beverly Hills, CA: Sage, 1972), 19-48.

17 Brady E. Hamilton, Joyce A. Martin, Michelle J.K. Osterman, and Lauren M. Rossen, "Births: Provisional Data for 2018," *U.S. Department of Health and Human Services*, Centers for Disease Control and Prevention, National Center for Health Statistics, National Vital Statistics

dropped to pre-2005 levels, undermining American population growth.[18] Decline in immigration to the U.S. was a major byproduct of the Trump administration pursuing a restrictive immigration agenda as a part of the Republican platform without due regard to economic ramifications.[19] This becomes even more problematic when considering that China's population is more than four times larger than America. The global race for talent should be considered a cornerstone of American strategy, and the fact that Canada surpassed the U.S. in 2021 as the top choice for foreign talent to attend university and work, should be considered an existential threat to long-term American growth and hard power.[20]

While the U.S. remains a leader in educational attainment, male high school graduation rates have declined—45,000 fewer male students graduate high school every year, making them ineligible for military or government service.[21] Approximately 84% of the nearly 1.3 million U.S. military service members are male.[22] As the available pool of military recruits continues to decline, the U.S. must contend with these realities and seek alternative paths to strategic power through smarter immigration policies. If these challenges remain unaddressed, it could undermine the economic vitality of the U.S. in the long-term, subsequently neutering its military power relative to Russia and China.

American grand strategy, as carried over from the Trump administration, ignores the historical antecedents that made America great in the first

System (NVSS) Report No. 007, May 2019, https://assets.documentcloud.org/documents/6003979/US-2018-Birth-Rate-Report-From-CDC.pdf.

18 "Unauthorized immigrant population trends for states, birth countries and regions," *Pew Research*, 12 June 2019, https://www.pewresearch.org/hispanic/interactives/unauthorized-trends/; Sabrina Tavernise, "Immigrant Population Growth in the U.S. Slows to a Trickle," *The New York Times*, 26 July 2019, https://www.nytimes.com/2019/09/26/us/census-immigration.html.

19 Sarah Pierce, *Immigration-related policy changes in the first two years of the Trump administration* (Washington, D.C.: Migration Policy Institute, 2019).

20 Roy Maurer, "Canada Replaces U.S. as Top Work Destination," *SHRM*, 8 April 2021, https://www.shrm.org/resourcesandtools/hr-topics/talent-acquisition/pages/canada-replaces-us-top-work-destination.aspx#:~:text=Canada%20is%20now%20the%20most,209%2C000%20people%20from%20190%20countries; Jeremy Neufeld, "STEM Immigration Is Critical to American National Security," *Institute for Progress* 30 March 2022, https://progress.institute/stem-immigration-is-critical-to-american-national-security/.

21 Richard Reeves, Eliana Buckner, and Ember Smith, "The Unreported Gender Gap in High School Graduation Rates," *Brookings*, 12 January 2021, https://www.brookings.edu/blog/up-front/2021/01/12/the-unreported-gender-gap-in-high-school-graduation-rates/.

22 Department of Defense, "*Department of Defense by Gender, Race and Ethnicity*," Office of Diversity, Management, and Equal Opportunity: Personnel Readiness, Pentagon, Washington, D.C. (2017). https://diversity.defense.gov/Portals/51/Documents/Presidential%20Memorandum/DoD%20Military%20by%20Gender%20Race%20and%20EthnicityV2.pdf?ver=2017-01-06-090352-110.

place, namely that the U.S. is a nation of immigrants.[23] The American ideal as a land of opportunity has remained an essential truth, enabling strategic dominance over near-peer adversaries. Valuing diversity and inclusivity beyond just immigration policies has practical ramifications; it is correlated with battlefield performance in modern war.[24]

With Russia's 2022 invasion of Ukraine, over 4 million educated Russians have fled the country, prompting the Biden administration to consider easing visa rules to capitalize on Russian brain drain.[25] Additionally, as of March 2021, international interest in attending American universities is back.[26] Thus, the rhetoric and actions of the Biden administration suggests that the 'soft power' image of the U.S. can be restored but only through the crafting of an explicit Biden administration *NSS* that values human capital and ties it to American economic progress and power.

The future of American hegemony rests on tapping into a 21st century *Strategic Brain Drain* policy against foreign countries through a well-crafted strategy of encouraging and fostering immigration to the U.S. in the context of great power competition.[27] This article begins by discussing the need to consider America's place in a globalized world, providing the impetus for an indirect approach and the need for valuing soft power and image and perception of the U.S. This lends itself to the Brain Drain policy argument for a liberalist approach to domestic and international politics, which ensures the U.S. can grow and maintain hard power at the expense of China and Russia.

---

23 Janice Fine and Daniel J. Tichenor, "A Movement Wrestling: American Labor's Enduring Struggle with Immigration, 1866–2007–Erratum/Corrigendum," *Studies in American Political Development* 23, no. 2 (2009): 218-248.

24 Jason Lyall, *Divided Armies: Inequality and Battlefield Performance in Modern War* (Princeton, NJ: Princeton University Press, 2020).

25 Zahra Tayeb, "The US plans to capitalize on a Russian 'brain drain' by easing visa requirements for workers with STEM skills, reports say," *Business Insider*, 30 April 2022, https://www.businessinsider.com/biden-plans-russian-brain-easing-visa-rules-stem-professionals-2022-4; Niko Vorobyov, "'Criminal adventure': Ukraine war fuels Russia's brain drain," *Al Jazeera*, 23 May 2022, https://www.aljazeera.com/news/2022/5/23/many-leave-russia-as-ukraine-war-drags-on.

26 Brendan O'Malley, "International students warming to US after Biden victory," *University World News*, 3 March 2021, https://www.universityworldnews.com/post.php?story=20210303133839873.

27 The "Strategic Brain Drain" concept was first introduced in a 2019 Strategic Multilayer Assessment (SMA) report. See: Jahara Matisek, "Outlasting China and Russia: An Alternative American Way to Victory in the 21st Century," in: *Power under Parity*, edited by Sarah Canna and George Popp, SMA Future of Great Power Competition & Conflict Project, J39, Pentagon, Washington, D.C. (September 2019): 12-14.

Five Objectives for the U.S. Operating in a Globalized World

Direct military competition with China and Russia poses major risks to global stability. One should take notice of the conclusions drawn by the likes of Adam Smith and Hans Morgenthau: economic power begets military power.[28] Investments and policies aimed at fostering development and economic progress, ensure national instruments of power are able to be fully wielded. However, pursuing American military power in the long-term as a primary strategy undermines the economic viability of the U.S. Thus, it is problematic that the U.S. spends half of its discretionary budget on national defense, while infrastructure and education remain underfunded, despite them being economic multipliers.[29] Instead, the pursuit of market efficiencies and technologies that keep America at the forefront of economic growth will naturally lead to military power, much as it did during the Cold War and before.[30]

Codifying the right blend of strategy, domestically and internationally, requires policymakers to first identify national interests and threats to them, It also requires placing a relative value on each of these, and whether certain problems can be identified as existential threats to the survival of America, or if these national interests can identified as vital, major, or peripheral.[31] Proper strategy also means considering what contributes to the power of a nation, and "measuring what matters," namely assessing assets and liabilities.[32]

Balancing domestic policy and foreign strategy is based on politicking by vested actors that can be manipulated by policymakers and senior government officials, including special interest groups such as American Israel Public Affairs Committee (AIPAC), thus driving Congress to provide billions of dollars of security aid to Israel with minimal conditions.[33] Each actor

28 Adam Smith, *The Wealth of Nations* (London: W. Strahan and T. Cadell, 1776), specifically Book V, Chapter 1; Hans Morgenthau, *Politics among Nations* (New York: Knopf, 1960). Morgenthau also mentioned the role of geography, to describe how the U.S. and the UK came to be formidable military powers.

29 Tatyana Palei, "Assessing the impact of infrastructure on economic growth and global competitiveness," *Procedia Economics and Finance* 23 (2015): 168-175.

30 Sonja Michelle Amadae, *Rationalizing capitalist democracy: The Cold War origins of rational choice liberalism* (Chicago: University of Chicago Press, 2003).

31 Donald E. Nuechterlein, "National interests and foreign policy: A conceptual framework for analysis and decision-making," *British Journal of International Studies* 2, no. 3 (1976): 246-266.

32 Michael Beckley, "The power of nations: Measuring what matters," *International Security* 43, no. 2 (2018): 7-44.

33 Historically, whenever the U.S. provides security aid, provisions are typically included that only American weapon systems can be purchased with this aid, and in the case of Israel, the country has been exempted from this requirement. For more on the power of special interest groups influencing American foreign policy, refer to: John J. Mearsheimer and Stephen M. Walt, *The Israel Lobby and US Foreign Policy* (New York: Macmillan, 2007).

has a competing self-interest in driving American grand strategy to elevate the intensity of a peripheral threat to one that supposedly poses an existential threat to the American way of life, and to promote their values as the key to combatting those threats. We have continued to see indicators of this in the past twenty years of U.S. armed conflict.

Intentional or not, post-9/11 era discourse continually elevated the terror threat emanating from Africa and the Middle East from a peripheral concern to an existential threat, contributing to the logic of sustained U.S. troop commitments in the region, much to Israel's chagrin.[34] Such rhetoric promoted and entrenched the U.S. in the Global War on Terror, despite Al Qaeda's existence as a terrorist organization intent on stoking fear rather than as a conventional force threatening to mass forces, land on American shores, and seize territory.[35] Indeed, political rhetoric has the ability to drive strategic decision making toward the perception and existence of an existential threat posed by adversaries lacking the power and ability to warrant this degree of concern. America's future strategic focus needs to evolve away from domestic politicking that overinflates the 'threat' of terror, leading to overfunding the fight against terrorism, and towards core concepts of generating U.S. power to out-compete rival powers. Shifting a kinetic strategy from the terror problem towards strategic competition does nothing other than refocus a new target with the same weapon systems. We need to refocus the target, but with different weapons systems. The evolving strategic competition with Russia and China needs to be a 'softer' competition of values and ideology where the U.S. offers a comparable advantage of better principals. Such a competing value proposition demonstrably outperforms illiberal, revisionist states.

For the U.S. to benefit from totalitarian politics in China and Russia, American leaders must better understand the complexities of the evolving international environment and integrate such nuanced thinking into the development of a domestic and foreign strategy to outcompete both. Capitalizing on Chinese and Russian authoritarianism requires an adjusted U.S. narrative publicly emphasizing a more palatable liberal ideology of inclusivity and cooperation. Emphasizing liberalism over realism on the world stage will produce an even more attractive citizen recruitment proposition for the best and brightest Chinese and Russian citizens to move to the United States for educational and work opportunities. This *Strategic Brain Drain* approach should be integrated into future U.S. grand strategy discussion. Adopting a strategy of intellectual attrition against adver-

34 Leonie Huddy and Stanley Feldman, "Americans respond politically to 9/11: Understanding the impact of the terrorist attacks and their aftermath," *American Psychologist* 66, no. 6 (2011): 455-467; Valentina Bartolucci, "Terrorism rhetoric under the Bush Administration: Discourses and effects," *Journal of Language and Politics* 11, no. 4 (2012): 562-582.

35 Richard Jackson, *Writing the war on terrorism: Language, politics and counter-terrorism* (Manchester, UK: Manchester University Press, 2018).

saries provides the U.S. with the necessary human and social capital needed to offset the population growth of China.[36] Such a 'talent capture' approach can galvanize U.S. innovation and rebalance the intellectual capital game for future strategic competition.[37] International perception matters too, meaning the fostering of a domestic political system and economy that is open to immigration and facilitates success. A complimentary system of reforms in American politics is needed ensure high levels of upward social mobility, keeping the perception of the "American Dream" alive for citizens and foreigners alike.[38] For instance, a clear pathway is needed for attaining U.S. citizenship after serving in the US Armed Forces.

Future *Strategic Brain Drain* success internationally, also requires domestic discourse that favors a liberal form of inclusive American nationalism, as some elements of U.S. society are trying to normalize illiberal, non-inclusive American nationalist identity.[39] It similarly requires a renewed social contract that emphasizes inclusivity and its value in American foreign policy.[40] For instance, a U.S. Army Colonel noted that "America has always believed that it represented its ideals through the lens of soft power .... Diversity is not decisive. It is another tool of power, and one that America possesses in abundance. It can be used better."[41] However, the waning of American soft power and the perception of it no longer being a welcoming nation of immigrants is at risk.

Soft Power Trumps Hard Power

The U.S. is still in a favorable position to outlast China and Russia in the 21st century, but only through indirect approaches. America is strategically and structurally poised to remain a hegemon, but only if U.S.

36 Michel Beine, Frédéric Docquier, and Hillel Rapoport, "Brain drain and economic growth: Theory and evidence," *Journal of Development Economics* 64, no. 1 (2001): 275-289; Ejiro U. Osiobe, "Human capital, capital stock formation, and economic growth: A panel granger causality analysis," *Journal of Economics and Business* 3, no. 2 (2020); Prasetyo, P. Eko, Andryan Setyadharma, and Nurjannah Rahayu Kistanti. "Social Capital: The main determinant of MSME entrepreneurship competitiveness." International Journal of Scientific & Technology Research 9, no. 03 (2020): 6627-6637.

37 Adam Tyson (ed.), *The political economy of brain drain and talent capture: Evidence from Malaysia and Singapore* (New York: Routledge, 2018).

38 Jennifer Wolak and David A.M. Peterson, "The dynamic American dream," *American Journal of Political Science* 64, no. 4 (2020): 968-981.

39 Jill Lepore, "A New Americanism: Why a Nation needs a National Story," *Foreign Affairs* (March/April 2019), https://www.foreignaffairs.com/articles/united-states/2019-02-05/new-americanism-nationalism-jill-lepore.

40 Jahara Matisek, Travis Robison, and Buddhika Jayamaha, "Extending the American Century: Revisiting the Social Contract," *Georgetown Journal of International Affairs* 20, no. 1 (2019): 5-15.

41 Mike Birmingham, "Diversity as power," *U.S. Army War College: War Room*, 18 May 2017, https://warroom.armywarcollege.edu/articles/diversity-as-power/.

leadership avoids counterproductive policies that elevate short-term gains in lieu of long-term payoffs. Policymakers need to elevate strategic 'soft' power decisions, such as continuing to be a beacon of democracy and promoting rule of law, inclusivity, diversity, and capitalism—attracting the most educated and talented people to America and retaining them. In cases of adversaries outmatching U.S. hard power, such issues can be more easily resolved through robust alliances and security partnerships.

Ideas about a vibrant, inclusive civil society require ways in which to foster notions of American identity and integration into norms and values of civic duty.[42] Strategic soft power in this case means retaining the foundations of a welcoming American society while respecting the rule of law and robust property rights. These are the fundamental norms, values, and institutions that Nobel Prize winning economist Douglas North attributed to being vital components of economic growth and development, to include foreign investment.[43] However, as evidence grows about how President Trump attempted to promote his "Big Lie" conspiracy of election fraud after losing, not to mention the failed 6 January 2021 insurrection, these recent events contribute to growing international perceptions of U.S. corruption and democratic backsliding.[44]

For the gender gap and declining population of qualifying males for military service, this means building on the foundation of the Women, Peace, and Security Act of 2017 by fully institutionalizing Defense Objective 1: that the "DoD exemplifies a diverse organization that allows for women's meaningful participation across the development, management, and employment of the joint force."[45] Women around the world are the fastest growing demographic obtaining higher degrees in Science, Technology, Engineering, and Math (STEM) and these skills are needed to compete in the global economy and defense sectors. Hence, attracting the best and brightest means making U.S. institutions attractive to women from both at home and abroad. Foreign engagement requires the shaping of the information environment in elevating the U.S. as valuing a liberal and inclusive society, while simultaneously high-

42 Lauren Gilbert, "Citizenship, Civic Virtue, and Immigrant Integration: The Enduring Power of Community-Based Norms," *Yale Law & Policy Review* 27 (Fall 2008): 335-398.

43 Douglas C. North, *Institutions, Institutional Change and Economic Performance* (New York: Cambridge University Press, 1990).

44 In 2016, the US was the 16th least corrupt country in the world, but by 2020 it fell to 25th least corrupt country, tied with Chile. Cailey Griffin and Amy Mackinnon, "Report: Corruption in U.S. at Worst Levels in Almost a Decade," Foreign Policy, 28 January 2021, https://foreignpolicy.com/2021/01/28/report-transparency-international-corruption-worst-decade-united-states/.

45 Terri Moon Cronk, "DOD Supports Women, Peace and Security Act, Official Says," DOD News, 27 July 2020, https://www.defense.gov/Explore/News/Article/Article/2288824/dod-supports-women-peace-and-security-act-official-says/source/GovDelivery/

lighting the perils of authoritarianism in China and Russia.

Success domestically and internationally, also requires people to know the rules of the game, so to speak. Every country has formal written laws, but the reality is that in most countries, including the United States, there are informal ways in which people coordinate economic activity and in authoritarian regimes, various activities are pursued without legal oversight and through corrupt negotiations with certain powerbrokers.[46] During the Cold War, the Soviet Union was bent on keeping the pace with a blistering U.S. nuclear weapons development program despite economic stagnation. However, the U.S. could afford such weapon building efforts because of the strong economy under the Reagan administration.[47] This is because the U.S. has been historically viewed as reliable when it comes to the rule of law and robust property rights, per the advocacy of Douglas North. North always emphasized these values, norms, and institutions as supporting economic stability, whereas countries that deviate from rules and laws discover that foreign investors lose confidence in the ability to do business in that country without the threat of corruption and rent-seekers undermining profits.[48] While strong authoritarian states can provide a stable-looking environment characterized by attractive investment opportunities, at least in the near term, long-term viability comes into question, because of certain regime expectations for companies to conform to illiberal policies (e.g., providing all data and information on customers to government officials). When confronted with the realities of anti-Western values and institutions compared to free market economies and democracy promotion as alternatives, long-term investors are more likely to favor the U.S. approach. The U.S. is a bastion of stability, meanwhile, dictators fail to follow international laws and norms and their countries suffer as a result of the perceived authoritarianism-based instability.

Strategically countering these illiberal states necessitates purposeful efforts through intelligence collection and media dissemination to display their malfeasance in governance. Such efforts are politically palatable and remain well-below threshold of conflict. Engaging in deliberate political and information shaping efforts would create and solidify the accurate impression of Russia being a state-sponsor of organized crime that is trying to *Russify* neighboring states, not to mention the numerous war-crimes being committed by Russia's pirate army in Ukraine.[49]

---

46 Douglass C. North, John Joseph Wallis, Steven B. Webb, and Barry R. Weingast (eds.), *In the Shadow of Violence: Politics, Economics, and the Problems of Development* (New York: Cambridge University Press, 2013).

47 Fred Chernoff, "Ending the Cold War: The Soviet retreat and the US military buildup," *International Affairs* 67, no. 1 (1991): 111-126.

48 Douglass C. North, *Institutions, Institutional Change and Economic Performance* (New York: Cambridge University Press, 1990).

49 Neil Hauer, "Putin's Plan to Russify the Caucasus: How Russia's New Language Law Could

Additionally, it is well documented that the Russian government works with the notorious Russian biker gang *Nochnye Volki* (Night Wolves) to subvert various Eastern European countries by espousing right-wing white nationalism.[50] Similarly important is changing the perception of China from the docile 'Panda' that it pretends to be, and highlighting China as a bellicose 'Dragon' trying to bully its neighbors in the South China Sea.[51]

China engages in subversive acts (e.g., stealing intellectual property [IP], weaponizing the supply chain, etc.), violates international laws in the South China Sea, and is increasingly reliant on a strategy of 'debt-trap diplomacy' in the underdeveloped world; seizing assets and infrastructure in countries that default on loan schedule paybacks.[52] For instance, while China appeared generous in building the African Union headquarters in Ethiopia, it was eventually discovered that China had 'bugged' the building for the purposes of cyber espionage.[53] Most importantly, not enough is being done by the West to paint the current danger of China, which is systematically eliminating its ethnic Uyghur Muslim population in favor of growing ethnic Han nationalism.[54]

Rather than pursuing a coordinated and costly military effort aimed at deterring Russian and Chinese aggression, especially in the gray zone, Western leaders need to showcase the illiberal tendencies of these strategic competitors.[55] The narrative of competition needs to be leveraged in favor of an alternative, long-term indirect strategy of attrition that will, in time, drain Russia and China of its social and human capital, leading to economic and military submission. Russian and Chinese deceit must rise to the forefront of the international conversation. The U.S. should avoid narratives that portend naked realism in favor of inclusive liberalist ideologies to strategically capture brain drain from China, Russia, and

---

Backfire," *Foreign Affairs*, 1 August 2018, https://www.foreignaffairs.com/articles/russia-fsu/2018-08-01/putins-plan-russify-caucasus.

50 Kira Harris, "Russia's Fifth Column: The Influence of the Night Wolves Motorcycle Club," *Studies in Conflict & Terrorism* (2018): 1-15.

51 Rob Gifford, "Panda-Huggers and Dragon-Slayers: How to View Modern China Today," *Social Education* 74, no. 1 (2010): 9-11.

52 Wilson VornDick, "Let China Fail in Africa," *The National Interest*, 29 January 2019, https://nationalinterest.org/feature/let-china-fail-africa-42812.

53 "African Union Bugged by China: Cyber Espionage as Evidence of Strategic Shifts," *Council on Foreign Relations*, 7 March 2018, https://www.cfr.org/blog/african-union-bugged-china-cyber-espionage-evidence-strategic-shifts.

54 Matthew Hill, David Campanale, and Joel Gunter, "'Their goal is to destroy everyone': Uighur camp detainees allege systematic rape," *BBC News*, 2 February 2021, https://www.bbc.com/news/world-asia-china-55794071.

55 Jahara W. Matisek, "Shades of gray deterrence: issues of fighting in the gray zone," *Journal of Strategic Security* 10, no. 3 (2017): 1-26.

other authoritarian states through both cultural and economic appeal. Thus, the U.S. and many other democracies have rightly played up strategic narratives about Russia's illegal and unjust invasion of Ukraine in 2022, as a way of isolating the country and undermining the perception of the country as a desirable place to work and live.[56]

The Russians are master manipulators and overly reliant on coercion, as seen with Russia's war against Ukraine.[57] Russia pursues various indirect strategies below the level of armed conflict against most of Europe without having to firing a shot. Russia 'misdirects' in Europe by bribing politicians, funding extremist political parties, sowing domestic dissent through information warfare, and worsening the COVID-19 pandemic by promoting anti-mask and anti-vaccine propaganda.[58] To some, the most effective capability Russia has is in its ability to marshal organized crime syndicates to act on their behalf to collect intelligence, smuggle, and eliminate rivals and dissenters abroad. Moreover, as described by one Swedish military intelligence officer interviewed, noted "Russia is weak," but mastered the employment of cheap "misdirection" at the expense of the U.S. and its allies in Europe.[59]

Meanwhile, Beijing targets Western civil society to create positive narratives about China, placing propaganda spewing Confucius Institutes at universities in the West and by requiring companies and entertainment industries (i.e. Hollywood, Disney) to abide by Chinese Communist Party (CCP) censors and rules to do business in China.[60] China engages in aggressive "Wolf Warrior Diplomacy," while building thousands of acres of artificial islands in the South China Sea for military purposes—disguising such actions as commercial fishing activity, and intimidating neighbors.[61] Worse, much like Russia, China undermines the global COVID-19 response by spreading "malign and subversive" information about the origins of the virus and ways of stopping its spread, further de-

56 Jill Goldenziel, "The Russia-Ukraine Information War Has More Fronts Than You Think," *Forbes*, 31 March 2022, https://www.forbes.com/sites/jillgoldenziel/2022/03/31/the-russia-ukraine-information-war-has-more-fronts-than-you-think/?sh=60583186a1e2.

57 Gunneriusson, Hakan, and Sascha Dov Bachmann, "Western Denial and Russian Control: How Russia's National Security Strategy Threatens a Western-Based Approach to Global Security, the Rule of Law and Globalization," *Polish Political Science Yearbook* 46, no. 1 (2017): 9-29.

58 Jahara Matisek and Buddhika Jayamaha, *Old and New Battlespaces: Society, Military Power, and War* (Boulder, CO: Lynne Rienner, 2022).

59 Interview, Swedish military officer, March 8, 2019.

60 Buddhika B. Jayamaha and Jahara Matisek, "Social Media Warriors: Leveraging a New Battlespace," *Parameters* 48, no. 4 (Winter 2018–19): 11-24.

61 Zhiqun Zhu, "Interpreting China's 'Wolf-Warrior Diplomacy,'" *The Diplomat* 15 (2020): 648-658; Gregory Poling, "Illuminating the South China Sea's Dark Fishing Fleets," *Center for Strategic and International Studies*, January 9, 2019, https://ocean.csis.org/spotlights/illuminating-the-south-china-seas-dark-fishing-fleets/

stabilizing Western governments and public trust in their institutions.[62] Such successful actions only gives more confidence and audacity to leaders in China to continue unabated, because they have scantly suffered any serious blowback from the West.

Owing to the strategic rebalance of the U.S. military, the pivot towards large scale combat operations to counter Russian and Chinese actions, seems to be focused on integrated deterrence and large-scale combat operation capabilities.[63] However, strategic soft power is a more advantageous approach given the structural realities facing China and Russia. For example, while China is expected to surpass the U.S. as the largest economy around 2030, this victory for Chinese will be fleeting. Lagging Chinese birth rates and an aging population will result in economic stagnation by 2040.[64] Moreover, China's shift towards tightening, authoritarian control of citizens (e.g., social credit rating score, etc.), and their recent moves to 'cleanse' and 'reeducate' Chinese Uyghurs, make China all the more vulnerable to civil strife and brain drain.[65]

In Russia's case, it is a dying country in terms of falling life expectancy rates, with one of the lowest birth rates in Europe.[66] Their revisionist leader Vladimir Putin attempts to exert influence for the purposes of making others *perceive* Russia as a great power. Hostile acts by Russia should be viewed through the lens of a country and people that once enjoyed rival status with America during the Cold War, but now has a shrinking economy that is smaller than Texas.[67] While Russia has attempted to build their own version of Silicon Valley in Skolkovo, this is a failing venture as the country lacks the necessary institutions, laws, and protections needed for the favorable conditions that are conducive to innovation and business growth. With Russia moving towards creating its own internet, much like China's "Great Firewall," this will only further reinforce authoritarian tendencies that will constrain civil society and institutions, undermining

62 Miriam Matthews, Katya Migacheva, and Ryan Andrew Brown, *Superspreaders of Malign and Subversive Information on COVID-19: Russian and Chinese Efforts Targeting the United States* (Santa Monica, CA: RAND, 2021).

63 Michael D. Lundy, "Meeting the Challenge of Large-Scale Combat Operations Today and Tomorrow," *Military Review* 98, no. 5 (2018): 111-118.

64 Yvan Guillemette and David Turner, *The Long View: Scenarios for the World Economy to 2060* (Paris: OECD Publishing, 2018).

65 Kate Lyons, "Uighur leaders warn China's actions could be 'precursors to genocide,'" *The Guardian*, December 6, 2018, https://www.theguardian.com/world/2018/dec/07/uighur-leaders-warn-chinas-actions-could-be-precursors-to-genocide.

66 Nicholas Eberstadt, "The Dying Bear-Russia's Demographic Disaster," *Foreign Affairs* 90 (2011): 95-108.

67 Frank Holmes, "Which Has the Bigger Economy: Texas or Russia?" *Forbes*, 17 April 2018, https://www.forbes.com/sites/greatspeculations/2018/04/17/which-has-the-bigger-economy-texas-or-russia/#52934a7670b9.

whatever economic growth and potential Russia has left.[68] What is the U.S. to do in exploiting these Chinese and Russian vulnerabilities? The most pragmatic approach are non-military solutions in the long-term, because it will ensure hard power can be generated when needed for mobilizing resources to confront an adversary.

The Brain Drain Policy

Jewish refugees escaping Nazi Germany in the 1930s "revolutionized U.S. science and technology."[69] Some of these refugees were vital members of the Manhattan Project, which contributed to the creation of the atomic bomb.[70] The Manhattan Project laid the foundation for American hegemony in the 20th century. A similarly inclusive American model will contribute to economic and military primacy in the 21st century and beyond.

The U.S. needs a *Strategic Brain Drain* policy to take advantage of the shift from the industrial age towards a globalized economy centered on information, service, and knowledge. As the 21st century moves away from industrial age economies, the quality of human capital will matter most in achieving efficiencies and markets of scale.[71] A post-industrial age economy dependent on digitized interconnectivity, rather than factories and manufacturing, requires both deliberate U.S. citizen recruitment and a new orientation to developing warfighting capabilities that gives the U.S. military a competitive edge.

Specific to citizen recruitment, the U.S. has a great comparable advantage with the internationally recognized perception of "The American Dream." The associated upward social mobility that a new immigrant in the U.S. might have by making a good living through hard work and dedication attracts talented individuals to the U.S. This American Dream— and the inclusive-liberalist ideology upon which it is built—must be a cornerstone of future U.S. grand strategy and leveraged in competition against totalitarian states. Beyond only social mobility, the American Dream allows for social acceptance of groups marginalized by the illiberal regimes of China and Russia. However, the "American Dream" is increasingly becoming a myth due to American upward social mobility ranking 27th

68 Scott Malcomson, *Splinternet: How geopolitics and commerce are fragmenting the World Wide Web* (New York: OR Books, 2016).

69 Petra Moser, Alessandra Voena, and Fabian Waldinger, "German Jewish émigrés and US invention," *American Economic Review* 104, no. 10 (2014): 3222-3255.

70 "Scientist Refugees and the Manhattan Project," *Atomic Heritage Foundation*, 20 June 2018, https://www.atomicheritage.org/article/scientist-refugees-and-manhattan-project.

71 Sunita Dodani and Ronald E. LaPorte, "Brain drain from developing countries: How can brain drain be converted into wisdom gain?" *Journal of the Royal Society of Medicine* 98, no. 11 (2005): 487-491; Jarand H. Aarhus and Tor G. Jakobsen, "Rewards of reforms: Can economic freedom and reforms in developing countries reduce the brain drain?" *International Area Studies Review* 22, no. 4 (2019): 327-347.

globally.[72] This downward trend, which began in 1980, should be considered a true existential threat to American hegemony in this century.[73] Political agreement and resolve, between both political parties in America, would be needed to generate policies that address wage stagnation and educational and economic opportunities to make the "American Dream" a reality again.[74]

Countries that pursue policies to exclude certain groups from participation in the economy, government, and military, are undercutting their own human and social capital. Promoting cultural equality and integration alongside policies that promote social mobility will allow the U.S. to recruit some of the most educated and talented individuals from authoritarian states, which increases the human capital of the American economy, and can result in technological advancements for the U.S. military.[75] Moreover, such a brain drain approach by the U.S. would make it difficult for these authoritarian states to be economically and militarily competitive, due to smaller pools of available human and social capital.

Aspirations by U.S. political and military leaders and think-tanks to always be exponentially more militarily powerful than hostile near-peer states is a cost-prohibitive strategy that will undermine economic growth in the long-term.[76] Military power is a relative concept with different value benchmarks to different factions. Whereas one faction argues military power is a reflection of total military force, another will argue it is, rather, a reflection of military capability. The resulting ambiguity of the concept continues to mire policy and law makers into a never-ending quest for more, in most cases simply to have more than the next. This is as wasteful as it is myopic. Adopting a numbers-based approach as a strategic guidepost to military power enhancement is steeped in illogical measures of power that have proven to be ineffective predictors of strategic success. A more productive "soft" power strategy focused on capturing *brain drain* is a better policy for undermining revisionist authoritarian states with hegemonic ambitions. American policies that rely on taking the world's best scientists, en-

---

72 Robert D. Putnam, *Our kids: The American dream in crisis* (New York: Simon and Schuster, 2016); "The Global Social Mobility Report 2020 Equality, Opportunity and a New Economic Imperative," *World Economic Forum*, January 2020.

73 Daniel Aaronson and Bhashkar Mazumder. "Intergenerational economic mobility in the United States, 1940 to 2000," *Journal of Human Resources* 43, no. 1 (2008): 139-172.

74 For examples and solutions, see: Ben Hecht, *Reclaiming the American dream: Proven solutions for creating economic opportunity for all* (Washington, D.C.: Brookings Institution Press, 2018).

75 Geoffrey D. Korff, "Reviving the forgotten American dream," *Penn State Law Review* 113, no. 2 (2008): 417-460; Richard Nadeau, Vincent Arel-Bundock, and Jean-François Daoust, "Satisfaction with Democracy and the American Dream," *The Journal of Politics* 81, no. 3 (2019): 1080-1084.

76 The most egregious advocate of America not spending enough on defense is the Heritage Foundation. "2021 Index of U.S. Military Strength: Executive Summary," *The Heritage Foundation*, 17 November 2020, https://www.heritage.org/military-strength/executive-summary.

gineers, doctors, and entrepreneurs is a needed strategy to have a comparative advantage in a globalized economy that will increasingly become transformed by the information era.[77]

Blending the Best of Realism and Liberalism?

American political leadership must craft a grand strategy that focuses on shifting and reshaping the public narrative to emphasize time-tested liberalism, with realism when necessary, such as current U.S. led efforts to punish Russia for its invasion of Ukraine. It is the necessary public narrative—the sales pitch—that sells the international community and sows the seeds of confidence into an American economic engine capable of maintaining global hegemony. Economic stability, more than military might, will be the foundation for success in great power competition and the one thing that will set the U.S. apart from its ambitious competitors. Despite their best efforts to cast themselves as democratic and capitalist, China and Russia are authoritarian and unrelenting regimes ripe for economic and social fracturing. The U.S. can generate trillion-dollar deficits with little effect on investor confidence as evidenced through the continued upward trend of the U.S. stock market through the COVID-19 pandemic. Likewise, ballooning spending seems to have little impact —at least for now—on the daily lives American citizens, in part because of America's concrete foundation and stability-promoting adherence to the rule of law and the resulting confidence it produces. Global powers and investors know this. China and Russia, meanwhile, are hollow economies presenting an illusion of growth and stability thinly veiled under revisionist and unsustainable ideologies. The U.S. has institutionalized relative economic stability; the Chinese and Russians have porous foundations susceptible to exploitation. Future U.S. grand strategy should promote economic power over military power, at least publicly.

The U.S. is a realist nation that advocates for a world of liberalism—as it has been since the end of World War II. But times are changing. The U.S. does not need or benefit from an overt "America First" policy platform as such a naked approach undermines U.S. efforts in the 21st century great power competition. Blatant advertising of *America First* is counterproductive; international image and perception increasingly matters in the internet age. This was most obvious in 2018, where a Pew Research Center study found that "America's international image continues to suffer."[78] A realist approach and

77 Peter A. Hall and David W. Soskice, *Varieties of capitalism: The institutional foundations of comparative advantage* (New York: Oxford University Press, 2001).

78 Richard Wike, Bruce Stokes, Jacob Poushter, Laura Silver, Janell Fetterolf, and Kat Devlin, "Trump's International Ratings Remain Low, Especially Among Key Allies: 1. America's International Image Continues to Suffer," *Pew Research Center: Global Attitudes and Trends*, 1 October 2018, https://www.pewresearch.org/global/2018/10/01/americas-international-image-continues-to-suffer/.

doing what is right for America, first, is a necessary front to ensure continued U.S. interests; boasting about it is not. Liberalism is the more palatable public rhetoric for the global stage. Unabashed realism can alienate American allies and fracture U.S. relations with contributing and influential nations in strategic competition whereas liberalism can unify.[79] This is not to say the U.S. should abandon realism or its *realpolitik*. The U.S. should take a page from E.H. Carr's playbook to reshape the narrative by injecting moralism back into Trump's *principled realism* of "America First."[80]

Conclusion

The U.S. can continue its realist approach under the umbrella of a liberalist label, benefitting itself first, and in doing so benefitting the rest. Confidence in the American model, and an inclusive liberal American identity, will attract outsiders who want to be a part of the U.S. economy. Foreign talent and faith in the American system, is what makes the U.S. economy so transformative and powerful. Absent a targeted recruitment strategy to attract talented global citizens, the U.S. is at risk of losing its hegemonic reserve of intellectual and entrepreneurial capital. The newest weapons and technological advancements may be dependent upon Chinese or Russian refugees. It could be fleeing Afghans, Uyghurs, or Chechens that might contribute to future development of artificial intelligence and quantum computing.

Those individuals seeking a better life may be the key to success for a country historically open to those willing to be a part of the greater American work ethic. Human capital translates directly into continued American economic and military power. Economic and military advances may be the lynchpin to offset competitors like China and Russia in the future of great power competition, and there is precedent for such an argument.

For American grand strategy to be successful in maintaining hegemony, it must remain reflective of the values the nation holds dear. International interest in attending American universities is back in 2021.[81] Additionally, the Biden Administration's softer approach to allies and partners, has immediately resulted in highly favorable views of the U.S. image.[82] Thus, the rhetoric and

79 Eliot A. Cohen, "America's Long Goodbye: The Real Crisis of the Trump Era," *Foreign Affairs* 98 (2019): 138-146.

80 E.H. Carr, *The Twenty Years' Crisis 1919-1939: An introduction to the study of International Relations*, 2nd edition (London: Macmillan, 1962 [1939]).

81 Brendan O'Malley, "International students warming to U.S. after Biden victory," *University World News*, 3 March 2021, https://www.universityworldnews.com/post.php?story=20210303133839873.

82 Richard Wike, Jacob Poushter, Laura Silver, Janell Fetterolf, and Mara Mordecai, "America's Image Abroad Rebounds with Transition from Trump to Biden," *Pew Research Center*, 10 June 2021, https://www.pewresearch.org/global/2021/06/10/americas-image-abroad-rebounds-with-transition-from-trump-to-biden/.

actions of the Biden administration suggests that the 'soft power' image of the U.S. can be restored. However, for America to truly benefit in the long-term, it means codifying a *Strategic Drain Brain* policy into national security documents and organizations. This can even translate into more effective negotiations with officials representing authoritarian governments, whereby safe-passage for the official and their family to America could be used as a tool for leverage to get information and/or to weaken an adversarial government.[83] It also means taking actions to attract and retain human and social capital, while also making domestic policy decisions that reduce corruption, advocate inclusive ideologies, and improve the "American Dream" of upward social mobility, ensuring that the worlds most talented decide to live in the United States. Achieving all of these objectives will only be possible if elected leaders on both sides of the political spectrum accept that the future of American economic and military power is dependent on innovation and entrepreneurship, which means fostering policies and reforming laws that promotes immigration to support American hegemony for the 21st century and beyond.

Ryan P. Burke holds a PhD from the Biden School of Public Policy and Administration, University of Delaware, and is a Full Professor in the department of Military and Strategic Studies at the U.S. Air Force Academy. A former U.S. Marine Corps officer, his recent book, *The Polar Pivot: Great Power Competition in the Arctic and Antarctica* (Lynne Rienner, 2022), illustrates how the Polar regions are becoming military domains in great power competition and how U.S. defense policy needs to adapt.

Jahara 'Franky' Matisek holds a PhD in Political Science from Northwestern University and is an active-duty U.S. Air Force officer and pilot assigned to the U.S. Naval War College as a Military Professor in the National Security Affairs department. His recent book, *Old and New Battlespaces: Society, Military Power, and War* (Lynne Rienner, 2022), demonstrates how the digital age is enabling everything to be weaponized as every citizen becomes a combatant.

---

83 Målfrid Braut-Hegghammer, "Cheater's Dilemma: Iraq, Weapons of Mass Destruction, and the Path to War," *International Security* 45, no. 1 (Summer 2020): 51-89.

Politics With Other Means: Aligning U.S. Military and Diplomatic Efforts in the Indo-Pacific Command Region

J. E. Schillo

Abstract

This paper offers potential answers to the question, "How does the United States effectively align military and diplomatic efforts to deter the People's Republic of China expansionism in the Indo-Pacific Command region?" A problem/solution method is used to address the question. The need for alignment is demonstrated by a thorough review of national-level policy documents, and the current gap in the alignment of military and diplomatic efforts is demonstrated through review and analysis of primary and secondary sources. An historical approach is taken throughout the work to provide background and context from Chinese and United States (U.S.) perspectives. An analysis is conducted to juxtapose eastern and western approaches to geopolitical competition and war. Historic examples of U.S. successes due to strategically aligned application of its Diplomatic, Informational, Military, and Economic (DIME) Instruments of National Power (IOP) are analyzed, as well as historic failures due to dis-alignment. Structural options which facilitate the alignment of military and diplomatic efforts within Indo-Pacific Command (INDOPACOM) are outlined and compared, resulting in the author's recommendation for implementation. Contrasting views are then addressed. Next, tailorable opportunities for global implementation of the recommended structural option are discussed. Lastly, an integrated Whole-of-Government (WOG) structural method is offered to support the oversight and achievement of national strategic objectives and provide the United States of America lasting and adaptable advantage across Great Power Competition (GPC) and beyond.

Keywords: Indo-Pacific, China, Geopolitical, Instruments of National Power, Great Power Competition

doi: 10.18278/gsis.7.1.6

Política por otros medios: alinear los esfuerzos militares y diplomáticos de EE. UU. en la región de comando del Indo-Pacífico

Resumen

Este documento ofrece respuestas potenciales a la pregunta: "¿Cómo alinea Estados Unidos de manera efectiva los esfuerzos militares y diplomáticos para disuadir el expansionismo de la República Popular China en la región del Comando del Indo-Pacífico?" Se utiliza un método de problema/solución para abordar la pregunta. La necesidad de alineación se demuestra mediante una revisión exhaustiva de los documentos de política a nivel nacional, y la brecha actual en la alineación de los esfuerzos militares y diplomáticos se demuestra mediante la revisión y el análisis de fuentes primarias y secundarias. Se adopta un enfoque histórico a lo largo del trabajo para proporcionar antecedentes y contexto desde las perspectivas de China y Estados Unidos (EE. UU.). Se lleva a cabo un análisis para yuxtaponer los enfoques oriental y occidental de la competencia geopolítica y la guerra. Se analizan ejemplos históricos de éxitos de EE. UU. debido a la aplicación estratégicamente alineada de sus Instrumentos de Poder Nacional (IOP) diplomáticos, informativos, militares y económicos (DIME), así como fallas históricas debido a la desalineación. Se describen y comparan las opciones estructurales que facilitan la alineación de los esfuerzos militares y diplomáticos dentro del Comando del Indo-Pacífico (INDOPACOM), lo que da como resultado la recomendación del autor para su implementación. Luego se abordan los puntos de vista contrastantes. A continuación, se analizan las oportunidades adaptables para la implementación global de la opción estructural recomendada. Por último, se ofrece un método estructural integrado de todo el gobierno (WOG) para respaldar la supervisión y el logro de los objetivos estratégicos nacionales y proporcionar a los Estados Unidos de América una ventaja duradera y adaptable a través de Great Power Competition (GPC) y más allá.

Palabras clave: Indo-Pacífico, China, Geopolítica, Instrumentos de poder nacional, Gran competencia de poder

其他方式实现的政治：协调美国在印太司令部地区的军事和外交活动

摘要

本文为“美国如何有效协调军事和外交活动以阻止中华人民共和国在印太司令部地区的扩张主义”这一问题提供了可能的答案。使用问题/解决法应对该问题。对国家级政策文件的全面审视证明了这种协调的必要性，同时，对原始资料和次级资料的审视和分析表明了当前军事和外交活动在协调方面的不足。本文通篇采用历史方法，从中国和美国的视角提供背景和情境。采用一项分析，将东西方对地缘政治竞争和战争采取的方法进行了比较。分析了一系列关于美国成功的历史实例，其中通过对外交、信息、军事和经济（DIME）这四种国家实力工具（IOP）的战略性协调应用而取得成功；并分析了因协调不一致而导致历史性失败的例子。概述和比较了促进印太司令部（INDOPACOM）内军事和外交活动协调一致的结构选择，从而得出作者的实施建议。随后描述了不同的观点。然后探讨了如何将结构选择适用于全球其他地区。最后，提出了一种综合的全政府（WOG）结构方法，用于支持国家战略目标的监督和实现，并为美国在大国竞争（GPC）及其他领域提供持久且适应性强的优势。

关键词：印度-太平洋，中国，地缘政治，国家实力工具，大国竞争

Clarification of Perspective and Terms. It is paramount to state upfront that the perspective of this work is one of respect, compassion, and empathy for the people of China. Any perceived critique of bureaucratic policy or action is not directed at the people of any nation, but rather intended to shape the reader's understanding of the current geopolitical dynamic. From the perspective of the United States and its partners and allies, the *Chinese Communist Party (CCP)*, in the form of the *People's Republic of China (PRC)*, may pose significant threats to national strategic objectives. However, the *people* of China are not a threat and should be valued and every effort should be made to secure and protect their individual and collective rights.

Here, the author uses the term *China* to refer to the commonly understood geographic region of Asia. It is also used when referring to claims of cultural identity, ethos, and achievement of groups collectively hailing from that same geographic region, spanning ancient times to the present. It is used rarely and in the context of quotations to refer to the current Chinese central government. The term *CCP* is used to refer to the specific Chinese Communist political party in China prior to the establishment of the Communist-led *PRC* in 1949. When refereeing to the post 1949 Chinese central government, the terms *CCP* and *PRC* are largely used interchangeably.

Introduction

Over the last two decades, while the United States has focused on the Global War on Terror and countering insurgencies in Iraq and Afghanistan, Communist China has risen. The PRC has seized every opportunity to gain global economic preeminence, assert itself as a regional hegemony, build an assertive posture toward the U.S., and increase its military capability and capacity to project greater power and influence globally (Jones 2020, 1). The PRC has attempted to isolate, manipulate, and maliciously influence U.S. partners and allies across Asia while taking advantage of the goodwill and opportunity provided by U.S. policies and actions (Cordesman, Burke, & Molot 2019, 59). The PRC's actions directly undermine the standing rules-based international order, threaten U.S. interests, challenge U.S. allies and partners, and place the PRC in direct GPC with the United States of America (Cordesman, Burke, & Molot 2019, 50, 53). The United States must effectively deter PRC regional expansion and aggression to ensure its national interests, maintain its geopolitical advantage, and support the international order and its Asian partners and allies.

The U.S. must take swift action to establish aligned organizational structure, focus, and policy implementation to deter PRC expansionism in the INDOPACOM region. At its disposal, the United States has DIME IOP to deter and counter the PRC's revisionist expansionism both regionally and globally (Henwood 2015, 2). This paper focuses regionally and asks the question "How does the United States effectively align military and diplomatic efforts to deter PRC expansionism in the INDOPACOM region?" Alignment of military and diplomatic IOP are key to achieving strategic objectives and drive follow-on informational and economic policies and actions to reach strategic end-states. Therefore, diplomatic and military IOP must be aligned first. Once this is achieved, informational and economic IOP can be aligned to enable diplomatic and military efforts regionally and globally. Analysis of current Department of Defense (DOD) and Department of State (DOS) policies will demonstrate an apparent gap in alignment and implementation of military and diplomatic efforts, even though the DOS's prioritized focus areas largely fall within traditional military areas of influence and responsibility (Department of State and U.S. Agency for International Development 2018, 24-25, 27).

If the U.S. does not align its military and diplomatic efforts now to counter and continue countering the

PRC's expansionism within INDOPACOM, it will rapidly lose regional influence, allies, security, and economic opportunity (Henwood 2015, 4). This outcome will degrade U.S. world standing, leave the U.S. vulnerable at home and abroad, and exacerbate economic challenges and political divisions within U.S. borders. First, this paper provides a comprehensive explanation of its analytical methodology and approach. Next, through a thorough literature review, this paper demonstrates the aforementioned alignment and implementation gap of U.S. military and diplomatic efforts across the DOD and DOS, and within INDOPACOM. Following that, a robust historical background of Chinese cultural identity, worldview, and interactions with the West is outlined to provide a baseline cultural context and perspective to the reader. Within this section, the rise of the CCP, establishment of the PRC, and U.S./PRC relations are outlined up to the current timeline to provide further context and perspective to better understand current expansionist efforts. Next, this paper analyses differences in eastern and western perspectives to provide illustrative examples of U.S. successes and failures due to IOP implementation and alignment and analyses the current state of U.S./PRC relations. This paper proposes and analyses two structural options for joint-interagency organizational constructs to effectively align military and diplomatic IOP within INDOPACOM and concludes with a recommendation. Implementation details of the recommended option are provided. Finally, future prospects for larger-scale implementation of the recommended joint-interagency structure across multiple Geographic Combatant Commands (GCCs) are discussed. Additionally, the author offers a potential joint-interagency global-level initiative which can enable the U.S. to gain and maintain geopolitical advantage worldwide through a WOG approach. This initiative enables regionally tailorable yet persistent alignment of all IOP across all GCCs to facilitate the achievement of strategic objectives and ensure the United States maintains geopolitical preeminence, supports its partners and allies, and gains and maintains a persistent advantage in Great Power Competition.

Methodology

This paper's analytical framework views the problem through the lens of GPC, leverages a problem/solution framework, maximizes a historical approach, and compares joint-interagency structural options to better align U.S. military and diplomatic IOP to deter Chinese expansionism within INDOPACOM.

To succeed in GPC, powers must be able to effectively operate in the "gray zone," influencing the actions of competitors and others below the threshold of armed conflict (Cordesman, Burke, & Molot 2019, 59). War is undesirable in a GPC construct; however, to effectively influence other actors with non-military IOP, Great Powers must maintain a credible military threat across the threshold of armed conflict to deter potentially malicious actors. If a credible military threat does not exist across that threshold, then most other IOP efforts can largely be ignored by competitors and adversaries since a hard-power enforcement capability does not exist. Additionally, Great Powers must be able to take limited-scale military action, if required, to influence below the threshold or war within a GPC construct.

It is understanding this reality within GPC considerations that this paper leverages a problem/solution framework to provide options and recommendations on ways to more effectively align U.S. military and diplomatic efforts to deter PRC expansionism in the INDOPACOM region. Problematic is the apparent gap in the alignment of DOD and DOS efforts to counter PRC expansionism within INDOPACOM, which is demonstrated in the literature review. The solution is the establishment and implementation of a regionally focused joint-interagency team tasked specifically with aligning DOD and DOS efforts to counter PRC expansionism within INDOPACOM. Two options for the formation of such a team are discussed, evaluated, and a final recommendation is made.

A historical perspective is provided throughout the background to provide the reader an understanding of China's self-identity from ancient times through the twentieth century and into the twenty-first century. Further context to the current U.S. / PRC GPC dynamic is provided by again leveraging a historical perspective to succinctly highlight key shifts in the relationship between China and the U.S. from the establishment of the PRC in the late 1940s through today. This is important because it highlights potential deltas in perspective, identity, and prime drivers between the PRC and the U.S., and offers insight into PRC's self-identity-based strategic objectives.

The importance of IOP alignment to achieve strategic objectives will be demonstrated many times over, driving home the point that alignment of military and diplomatic efforts enables coordinated strategic approaches. Which would in turn enable synchronized governmental operations and mutually supporting tactical-level engagements, resulting in a united DOD/DOS front within INDOPACOM to deter PRC expansionism. This analysis relies heavily on a qualitative approach but touches on quantitative metrics as well. The overall approach leverages a historical perspective to provide the reader examples of effective U.S. military/diplomatic alignment and the positive resultant outcomes. Additionally, key points where the U.S. has experienced setbacks and failures in international relations due to the dis-alignment of military and diplomatic efforts are highlighted.

Based on thorough analysis, two structural organizational options are offered, and associated recommendations are provided to more effectively align U.S. military and diplomatic efforts within INDOPACOM to counter PRC expansionism within INDOPACOM. These two options are compared/contrasted via a Course-of-Action-like comparison method tied to criteria outlined in that section. Based on this comparison, a joint-interagency organizational option is recommended. The future prospects and conclusion portions demonstrate how this model could potentially be implemented as structural blueprints for similar IOP alignment across multiple State Department Regions/Bureaus and GCCs.

Literature Review—Demonstrating the Gap

This paper offers options which enable the U.S. Government to more effectively align military and diplomatic efforts to deter Chinese expansionism in the INDOPACOM region. To do so, the gap in alignment must be identified, addressed, and plans put in place to coordinate, synchronize, and align military and diplomatic efforts within INDOPACOM. The primary and secondary sources reviewed below demonstrate the existence of a gap and the validate the need for alignment in military and diplomatic strategic IOP alignment.

U.S./PRC GPC has been and remains a primary focus of the U.S. Office of the President. In his *Interim National Security Strategic Guidance* issued in March 2021, President Biden specifically addresses the PRC's aggressive actions and the U.S.'s duty to hold them accountable (Biden 2021, 10). President Biden specifically states: "I direct departments and agencies to align their actions with this guidance," which also implies the need for departments writ large to align strategic efforts laterally, as well as vertically (Biden 2021, introduction). President Biden stresses "diplomacy as our tool of first resort" while ensuring "responsible use of our military," with the strategic objective of enabling the U.S. to "prevail in strategic competition with China" (Biden 2021, 12, 20). Even though this is interim guidance until the Administration publishes a new National Security Strategy (NSS), the need for strategic alignment across government departments and agencies and the currently existing gap in that alignment is clear.

Previous U.S. Presidents also recognized and prioritized U.S. GPC with the PRC. In the *National Security Strategy of the United States of America 2017* (2017 NSS), President Trump outlined priorities and engagement strategies with the PRC across the GPC continuum. GPC exists on a scale and ranges from cooperation to conflict, but rather maintaining U.S. leadership in the region to ensure markets and trade agreements remain open and profitable for U.S. economic endeavors and interests were top priorities of his administration as well. To effectively achieve these goals, the U.S. must maintain its international allies and partners across INDOPACOM. President Trump articulated that PRC military and economic expansionism are unfairly threatening the U.S. and regional nations' markets shares and specifically called on allied and partnered nations to support the U.S.-led efforts to counter the PRC in the political/diplomatic arena, the economic sector, and across the military and security sectors (Trump 2107, 45-46). The implied task in this strategy is to align efforts across the instruments of national power to dis-allow and deter PRC expansionism.

Secretary of Defense (SecDef) Lloyd J. Austin specifically states the need for a "coordinated and synchronized" WOG approach; he has discussed his vision for the DOD to prioritize the PRC as the "number one pacing challenge" today (Austin 2021, 1). This statement is telling in that it clearly articulates the need for alignment, which does not currently exist. The SecDef's vision of strategic IOP alignment and a WOG approach, where the military is *an* IOP and not *the* IOP, is indicative of the need for balanced and aligned strategic approaches across all IOP. This demonstrates the existing gap for which this paper provides analysis and corrective recommendations.

In the *Summary of the 2018 National Defense of Strategy of The United States of America: Sharpening the American Military's Competitive Edge*, former SecDef James Mattis outlined his assessment of the competitive strategic environment and DOD objectives, nested in President Trump's *2017 National Security Strategy*. He opens by stating that the DOD must reinforce "America's traditional tools of diplomacy," and even though the PRC "seeks Indo-Pacific regional hegemony in the near-term and displacement of the United States to achieve global preeminence in the future. The most far-reaching objective of this defense strategy is to set the military relationship between our two countries on a path of transparency and non-aggression" (Mattis 2018, 1-2). Even though military and diplomatic alignment is an implied vice specified task in this statement, it is still very clear that the former SecDef intended to facilitate transparency and avoid escalations to the point of conflict. This can be best facilitated by the alignment of U.S. diplomatic and military IOP to enable effective strategic policy implementation in GPC with the PRC.

Nested under the *2018 National Defense Strategy*, the Joint Staff's *Description of the National Military Strategy 2018* provided the Chairman's of the Joint Chiefs of Staff "military advice for how the Joint Force implements the defense objectives in the NDS and the direction from the President and the Secretary of Defense" (The Joint Staff 2018, 1). This document acknowledges GPC with the PRC across all domains, highlights the need for allies and partners, and specifically outlines "Compete Below the Level of Armed Conflict (With a Military Dimension)" as a Mission Area, but stops short of specially stating a need for aligned military and diplomatic efforts (The Joint Staff 2018, 2-3). However, to operate successfully in GPC against a peer competitor, alignment of military and diplomatic efforts is critical. DOD/DOS alignment is an implied task in this document and remains a gap writ large.

The *Joint Strategic Plan FY 2018-2022* outlines the collaboration between DOS/U.S. Agency for International Development (USAID) and the DOD in a few general categories but stops short of outlining a need for aligned objectives (Department of State and U.S. Agency for International Development 2018, 24, 27, 29, 31, 60). Ironically, the first three DOS/USAID strategic objectives in the diplomatic *Joint Strategic Plan* are military-centric strategic objectives: "Strategic Objective 1.1: Counter the Proliferation of Weapons of Mass Destruction (WMD) and their Delivery System"; "Strategic Objective 1.2: Defeat ISIS, al-Qa'ida and other transnational terrorist organizations, and counter state-sponsored, regional, and local terrorist groups that threaten U.S. national security interests"; "Strategic Objective 1.3: Counter instability, transnational crime, and violence that threaten U.S. interests by strengthening citizen-responsive governance, security, democracy, human rights, and the rule of law" (Department of State and the U.S. Agency for International Development 2018, 24, 25, 27). The cross-boundary nature of these objectives further demonstrates the need for specified alignment between military and diplomatic IOP across the DOS and DOS writ large, specifically concerning the PRC's expansionist efforts in the INDOPACOM region.

In his *Integrated Country Strategy - China,* the U.S. State Department Chief of Mission (COM) for China outlines mission priorities, strategic framework, mission goals and objectives, and management goals and objectives. The COM points out that even though we work in cooperation with the PRC on many fronts, our "core values contrast sharply," resulting in conflicting interests (Chief of Mission 2018, 2). The COM outlines balanced trade as a top priority while denying PRC attempts to illegally or maliciously acquire foreign or proprietary technology (Chief of Mission 2018, 2). With this baseline established upfront, the COM goes on to outline cooperative activities between the U.S. and the PRC which adhere to trade law and international norms. This DOS regional strategy does not mention U.S. military presence or efforts in the region, let alone a synchronized or aligned approach between the DOS Asian Bureaus or the INDOPACOM GCC. This continues to illustrate the gap in the alignment of strategy, plans, and implementation.

The Congressional Research Services report *Renewed Great Power Competition: Implications for Defense - Issues for Congress,* outlines assessed national defense issues and focus areas that will be affected by GPC. The report outlines the military drawdown post-Cold War, but given its comparison of the current situation to the Cold War, one could infer that the report is advocating for an increase in military size to effectively compete with other Great Powers; however, the report seems to stop short of making any overt recommendations (Congressional Research Services 2020). Again, no analysis is offered on how to align military and diplomatic operations as part of a holistic strategy.

Bruce Jones' *China and the Return of Great Power Competition* offers a strategy for countering Chinese expansionism through military posturing, building alliances, or various types of deterrence. Jones even goes as far as suggesting a type of dual-globalization factions, one led by the U.S. and the other by the PRC (Jones 2020, 9). However, Jones does not touch on how to align military and diplomatic efforts in his article, further demonstrating how this gap is largely overlooked when developing strategic approaches or translating them into operational coordination and tactical actions.

Jeffery Bader, former principal advisor to President Obama on U.S. policy in Asia, and now a senior fellow in the John L. Thornton China Center at the Brookings Institution continues to state the need for aligned strategic approaches and multi-lateral engagement, in *U.S.- China Relations: Is It Time to End the Engagement?* Bader advocates for multi-lateral international engagement with the PRC to open markets and ensure equity and obligations in the marketplace (Bader 2018, 5-6). This type of engagement would require a WOG approach on the part of the U.S. to gain other nations' support across this multilateral engagement, but Bader states that the U.S. must "put our own house in order" first (Bader 2018, 6). This requires strategic, operational, and tactical alignment of our military and diplomatic efforts within INDOPACOM, which currently does not exist at the level required for the U.S. to gain and maintain the advantage in a GPC construct with the PRC.

Across these sources, it becomes evident there is a need for strategic alignment of U.S. IOP. The implied task of IOP alignment across U.S. governmental departments is present in guidance and directives from current and former U.S. Presidents, both of whom recognize that the U.S. is operating within a GPC construct with the PRC. These realities are echoed by current and former Secretaries of Defense, and within DOS/USAID Strategic Objectives, which mirror military objectives—further driving home the point of needed alignment (Department of State and US Agency for International Development 2018, 24, 25, 27). Refining focus into INDOPACOM, the gap in the alignment of efforts across military and diplomatic entities at the operational and tactical levels becomes even more glaring when examining the DOS *Integrated Country Strategy – China*. Every one of the sources reviewed, either explicitly or implicitly, demonstrates the gap in the alignment of military and diplomatic IOP at every level. If this gap is lessened by more effectively aligning military and diplomatic efforts based on the recommendations of this paper, then the U.S. can more effectively deter PRC expansionism within INDOPACOM.

Background

The Art of War is a seminal work attributed to the legendary Sun Tzu, which compiles ancient Eastern thoughts and perspectives on warfare strategy and philosophy. By understanding key tenets outlined therein, one may begin to understand prime motivators and drivers in the PRC's strategic approaches to GPC today. As articulated in *The Art of War*, to succeed one must know both their adversaries and oneself (Sun Tzu 2000, 11). Keeping this in mind, the following background provides a deeper understating of China's historical identity; its influence across Asia and beyond; its past interactions with the western world; U.S./PRC relations from 1949-2000; and U.S./PRC relations from 2001-present. The historic approach sets the stage for effective analysis and provides a baseline from which to better understand the PRC's viewpoints and perspectives to more effectively leverage and align IOP to gain desired strategic outcomes.

China's Historical Identity

China's historical and cultural identities today are rooted in thousands of years of empire, innovation, and influence. China is Earth's oldest continuous civilization, and oral tradition tells of Emperor Yu establishing the Xia Dynasty which lasted from approximately 2100-1600 B.C. (History.com 2019, China: Timeline). This period was followed by the Shang Dynasty, from approximately 1600-1050 B.C., when advances in mathematics and astronomy marked the collective advancement of their people (History.com 2019, China: Timeline). Ancient Chinese history and oral tradition boast vast rising empires and advances in science and technology contemporary with the rise of ancient Babylonian and Egyptian dynasties in the Middle East. From 551-479 B.C., the globally renowned philosopher Confucius lived and worked across China developing schools, disciples, and concepts that are still central to the Chinese people and government today (History.com 2019, China: Timeline). The Qin Dynasty unified disparate warrior-ruled territories into what we now call China, and in the 200s B.C., standardized national scripts established an imperial academy, created the 500-mile long Strait Road, began the Great Wall, and built vast underground chambers and the associated famed terracotta warrior statues (History.com 2019, China: Timeline).

The Han Dynasty followed the Qin in the second century B.C. The Han Dynasty and its achievements became and remain the cultural touchpoint for Chinese self-identity and a key reference point of cultural heritage. The Han Dynasty established the Silk Road connecting East and West trade routes and promulgating wealth and influence from central Asia across the continent and into the Middle East—China's first Golden Age (History.com 2019, China: Timeline). At the beginning of the second century A.D., the Han developed paper and spread it across the empire resulting in the first Chinese dictionary and the first book of Chinese history (History.com 2019, China: Timeline). Modern China is aware of and confident in their ancient history and Han cultural identity, and currently uses these and other historic reference points across Han and other dynasties as motivating factors to drive a return to preeminence as a global Great Power in the 21st century.

In the later centuries A.D., China continued to lead the way in technology, scientific innovation, and empire-driven governmental structure. In the ninth century A.D., a Second Golden Age under the Tang Dynasty brought forth gunpowder and the development of the first printing press in China, both of which are watershed inventions and were tipping points in human history (History.com 2019, China: Timeline). These achievements harkened back to Chinese greatness first experienced under the Han and reinforced traditional Han cultural identity across China. Across the next few centuries, the Tang Dynasty experienced periods of upheaval and limited stability. Specifically, the An Lushan Rebellion in the eighth century A.D. opened the door to the eventual downfall of the Tang Dynasty (Encyclopedia Britannica online, n.d., An Lushan Rebellion summary). This internal rebellion against the Tang Emperor, led by An Lushan, one of his own generals of non-Chinese origin, was eventually quelled but the Tang were greatly weakened and the door opened for following power grabs (Encyclopedia Britannica online, n.d., An Lushan Rebellion summary). In the centuries to come, China became fragmented by warlords and the rival overlapping dynasties of the Song, Liao, and Jin collectively spanning from the early 900s A.D. to roughly the mid-1200s A.D. (Encyclopedia Britannica online, n.d., The Song (960–1279), Liao (907–1125), and Jin (1115–1234) dynasties). Eventually these kingdoms gave way to the Mongol conquest of Genghis Khan across much of Asia; and Genghis' grandson Kublai Khan unified Mongolia and China, and whose influence spread to parts of the Middle East, Europe, and Siberia under the Chinese Yuan Dynasty (History.com 2019, China: Timeline).

Post-Yuan Dynasty, China's first attempt to build a large blue water navy emerged under the Ming Dynasty and resulting in historic territorial claims to the South China Sea, which are still leveraged today (Gronewold, n.d.). The Ming Dynasty came to power in the mid-to-late 1300s A.D. and ruled China until northern Manchurian incursions resulted in the establishment of the Qing Dynasty in the late 1600s A.D. (Gronewold, n.d.). The Ming Emperor Zhu Di (the Yongle Emperor) desired to distinguish the Ming from the previous Mongolian Yuan rulers and ordered great sea expeditions to demonstrate Ming power (Gronewold, n.d.). Under the command of Admiral Zheng He, Ming "Treasure Ships" sailed from China to Southeast Asia, establishing Chinese hegemony across those regions and forming the basis for Chinese territorial claims still made today (Gronewold, n.d.).

Zheng He sailed on to India, the Persian Gulf, and Africa establishing nautical trade routes and outposts along the way (Gronewold, n.d.). For perspective, many of Zheng He's "Treasure Ships" were "sailing vessels over 400 hundred feet long, 160 feet wide, with several stories, nine masts and twelve sails, and luxurious staterooms complete with balconies. The likes of these ships had never before been seen in the world, and it would not be until World War I that such an armada would be assembled again" (Gronewold, n.d.). Contemporary European vessels paled in size and scope when compared to these feats of Chinese nautical engineering. These expeditions were comprised of tens of thousands of sailors and established China as the regional Great Power across Asia (Gronewold, n.d.). Not only did the Ming spread Chinese influence across the eastern hemisphere and establish self-perceived cultural precedence for China as the regional hegemony, but they established historical roots for Chinese identity as a powerful sea-faring naval culture.

This again harkened back to the roots of historic Han cultural identity but also opened a new area, the South China Sea, for the reestablishment of a Chinese golden age. However, in 1557 A.D. the Ming Dynasty allowed a European trade presence in China, which changed the trajectory of the Asian continent, and Chinese power, forever (History.com 2019, China: Timeline).

China and Western Powers

Upon establishing permanent contact with the west, one could assert that China has experienced a relatively continuous wave of conflict and perceived exploitation. In 1683 A.D., the Ming Dynasty seized Dutch-controlled Taiwan, which was later claimed by the Manchurian Qing Dynasty once they supplanted the Ming (History.com 2019, China: Timeline). This dynamic was the beginning of the struggle over the control of Taiwan we still see driving geopolitical controversy today. This dynamic also continued to perpetuate the rise-and-fall cycle of Han cultural identity based on northern invasions and started a tumultuous relationship with the west. All of these factors shaped and continue to shape the Chinese perspective, objectives, and geopolitical actions today.

The Manchu Qing were invaders from the north, who encroached on the Han cultural identity of most Chinese people and disallowed the reestablishment of a New Chinese Golden Age under the Ming. Throughout Chinese history a repeated pattern occurs where Han cultural identity and leadership results in innovation and rise to Great Power status, only to be setback by invaders. Within this setback, the people of China harken back to Han identity and self-perceived destiny to reestablish regional power and hegemony. The rise of the Qing Dynasty in the mid-1600s is the invading/setback portion of this cycle. As will become self-evident upon further historical discussion, today China is attempting to rise again to New Golden Age in this cycle.

Under the Qing Dynasty, the United States first established loose diplomatic relations with China in 1784, but trade was largely dominated by European nations (Council on Foreign Relations, n.d.). In the 1830s, American Protestant missionaries arrived in China; by the mid to late 1840s, the U.S. and China had signed trade agreements and Chinese immigrant laborers began to travel to the U.S. (Office of the Historian, n.d.). Earlier in that same decade, the British Empire perpetuated an opium addiction crisis among the people of China to develop a market for Indian opium and gain advantage over the Qing Dynasty by disrupting their population base through drug addiction, which resulted in the Opium Wars and British control of Hong Kong through the end of the twentieth century (History.com 2019, China: Timeline).

The Taiping Rebellion, led by a Chinese self-proclaimed Christian prophet, took 20 million lives and challenged Qing control from the 1850s through the mid-1860s (History.com 2019, China: Timeline). This instability was largely perceived by the ruling Qing to be a direct result of contact with the west and western religion. The Taiping Rebellion was followed by the Second Opium War during which Britain and France demanded the legalization of opium which had been outlawed due to its atrocious effects on the population during the First Opium War (History.com 2019, China: Timeline). This war resulted in the ceding of more economic and territorial power by the Chinese to the Europeans (History.com 2019, China: Timeline).

Due largely to its loss of control of its own historically held territories to the Europeans, China's regional hegemony began to be challenged. The Japanese gained control of Taiwan in the First Sino-Japanese War, further degrading China's regional status (History.com 2019, China: Timeline). Due to concerns over potential geopolitical fragmentation in China due to western nations' overreach, in 1899 and 1900 U.S. Secretary of State John Hay, with the intent of bolstering Chinese stability and commerce, issued the Open Doors Notes to other foreign powers attempting to claim exclusive access to traditional Chinese spheres of influence (Council on Foreign Relations, n.d).

As a result of western interactions, at the turn of the century, the internal Boxer Rebellion strove to oust western influence from China and drew combined military action from an eight-nation western coalition when the peasant rebels began attacking and killing Chinese Christians and Christian missionaries in response to a perceived threat to Chinese cultural identity (Plante 1999). This fragmented a previously unified Chinese governmental policy and split loyalties when Chinese "Authorities sometimes fought to protect foreigners and Christians and at other times chose to do nothing at all" (Plante 1999). By 1901, a formal peace treaty was signed, which favored western interests (Plante 1999).

China in the Early-to-Mid Twentieth Century

The factors which led to the rise of the CCP, the war between Chinese Communists and Nationalists, and the establishment of the PRC are many and complex and could singularly consume years of research and many volumes of text. Even though they resulted in the current form of government in China, they are still only a small portion of Chinese history and identity, and seem to be part of a larger overall cycle in the region. The intent of the brief coverage of these events within this paper is not to downplay their many factors or significance, but to offer a concise snapshot of their emergence which has led to the current state of play in U.S./PRC GPC.

Turmoil continued to emanate across China, which Communist scholars have ascribed to western ideas and cultural contact. Western-educated Sun

Yat-Sen led the Xinhin Revolution of 1911 and the following Wuchang Uprising where 15 provinces revolted against the Qing Dynasty (History.com 2019, China: Timeline). Sun took control of the government and declared The Republic of China in 1912 (History.com 2019, China: Timeline). Continued social and cultural unrest resulted in the establishment of the Chinese Communist Party in 1921 (History.com 2019, China: Timeline). The Chinese Communists endured the brutal Shanghai Massacre in 1927 at the hands of Chinese Nationalist Leader Chiang Kai-Shek, which eventually resulted in the Chinese Civil War from 1931 to 1949 (History.com 2019, China: Timeline).

During World War II, somewhat of an internal respite occurred between 1937-1945, when the CCP and the Nationalists formed a tense alliance against the Japanese called the Second United Front—the first having been from 1924-1927 when Sun Yat-sen allowed CCP members to ally with his Nationalists in exchange for Soviet military aide (Encyclopedia Britannica online, n.d., United Front; Chinese history [1937-1945]). Throughout World War II, the United States backed Chiang and his Chinese Nationalists against invading Japanese forces (Council on Foreign Relations, n.d.). However, in 1949, the CCP gained the advantage and Chiang and many of his Nationalist soldiers were exiled to Taiwan, where they continued to offer dissent and dispute the claims of the CCP. The CCP emerged victorious from the Civil War and Mao Zedong arose as the leader of the People's Republic of China, officially established in 1949 (History.com 2019, China: Timeline).

Even though no strangers to perpetrating atrocities on their own people to gain and maintain political control, from both a power-based and socio-cultural perspective the PRC largely views western interactions with China as negative. The PRC views the western world order as a challenge to historic Chinese power claims. The PRC perceives historic western economic interactions as exploitation of the Chinese people and views western culture as a corruption to Chinese culture. The PRC's relationship with the U.S. began after the U.S. had already backed their enemy, Chiang Kai-Shek. In their view, the establishment of the PRC returned control of China to the Chinese people, if in name only. It is through these lenses that the PRC began its post-World War II relationship with the western world, especially with the United States of America. This fed the PRC's self-perception of heroically ending foreign control of China, perpetuated sociological reach-back to Han cultural identity, and commenced the portion of the historical cycle where Han culture can begin rebuilding what the invaders have setback.

U.S./PRC Relations 1949-2000

Due to all of the factors outlined above, U.S./PRC relations were limited and cold from the start. Continued U.S. support to Chinese Nationalists, then occupying Taiwan, and the standing global geopolitical schism between nations claiming Capitalist or Communist economic systems exacerbated already tense yet practically non-existent international relations (Council on Foreign Relations, n.d.). Relations were further chilled by the Korean War, where the Chinese and Soviet Communist governments backed communist North Korea and the U.S. fought alongside the South Korean, the Republic of Korea (ROK), forces. The 1953 Armistice Agreement halted open combat actions in the Korean War, though technically the war still has not ended; however, the following year the removal of U.S. naval blockades allowed Chiang Kai-Shek and his Nationalist troops to launch from Taiwan and seize disputed islands, causing the first Taiwan Strait Crisis (Council on Foreign Relations, n.d.). The PRC responded by indirect fire attacks on the seized islands, which drove the U.S. and Chiang into a mutual defense treaty (Council on Foreign Relations, n.d.). Hostilities ceased and Nationalist forces withdrew only after the U.S. threatened a nuclear attack on the PRC (Council on Foreign Relations, n.d.).

Mao Zedong then began to focus internally on perceived threats to his power within the borders of the PRC. From 1958-1962, Mao's "Great Leap Forward" attempted to force industrial revolution across an agrarian-based society by mandating PRC-owned farming communes and outlawing private farming, resulting in nation-wide famine and approximately 59 million deaths (History.com 2019, China: Timeline). Mao continued to tighten his grip with the 1966 "Cultural Revolution," which intentionally removed traditional Chinese cultural influence and any Capitalist undertones from society (History.com 2019, China: Timeline). This totalitarian crackdown began with Maoist ideological indoctrination in schools and the formation of units of school-aged Red Guards to violently attack "undesirable citizens" (History.com 2019, China: Timeline). However, it rapidly evolved into fear-based implementation of martial law, and CCP-wide internal purges resulting in approximately 1.5 million deaths within the PRC (History.com 2019, China: Timeline). These destructive policies, violet actions, and campaigns of terror against the Chinese people by their own government left them with nothing but fear of and distrust in that same government, even if they complied with its policies. From an international relations standpoint, Mao's actions further isolated the PRC from international engagements and exponentially increased tension across U.S./PRC relations.

Relations remained on razor's edge until 1969 when a Sino-Soviet border dispute resulted in a re-opening of U.S./PRC relations (Council on Foreign Relations, n.d.). This was followed by an invitation from the Chinese National Ping-Pong Team to the U.S. National Ping-Pong Team to visit China in 1971, as well as a secret visit that same year by Secretary of State Henry Kissinger to begin opening internal relations (Lu, n.d.). As relations were gradually established, President Nixon visited the PRC in 1972 and signed the Shanghai Communique to establish an official geopolitical dialog, specifically concerning Taiwan, with the PRC (Lu, n.d.).

Relations slowly progressed over the next decade. However, confusion resulted in 1979 when President Carter officially recognized the PRC diplomatically, recognized their "One China" policy, and cut ties with Taiwan; but that same year the U.S. Congress passed the Taiwan Relations Act to enable continued U.S./Taiwan relations and U.S. arms support to Taiwan (Council on Foreign Relations, n.d.).

Shortly thereafter, President Reagan's administration balanced U.S. interactions between the PRC and Taiwan in efforts to maintain stability in the region while countering Soviet expansionism. This resulted in "Six Assurances" to Taiwan, and a Third Joint Communique with Beijing, allowing U.S. arms sales to PRC (Council on Foreign Relations, n.d.). The Reagan Administration's diplomatic efforts with the PRC opened the door to the U.S. market, enabled increased U.S. trade, and helped kick start the Chinese economy. Throughout the 1980s and 1990s, the PRC's approach to internal domestic policy, human rights, and free speech were challenged and at times condemned by the West and the U.S. in particular. However, due to Reagan's diplomatic outreach and special economic policies with the PRC, the nation developed an industrial base which grew into a manufacturing and economic powerhouse. However, the CCP violently resisted efforts from their own members to work towards a more democratic society, specifically demonstrated by policies of suppressing free speech and the tragic events of the Tiananmen Square Massacre in 1989 (History.com 2020, Tiananmen Square Protests). Chinese President Jiang Zemin and U.S. President Bill Clinton and both conducted diplomatic visits to each other's nations in 1997 and 1998, respectively (Lu, n.d.). By the late 1990s, U.S./PRC relations appeared to be on a positive glide slope.

In 2000, the PRC and the U.S. officially normalized trade relations and the PRC joined the World Trade Organization (WTO) in 2001 (Council on Foreign Relations, n.d.). While the U.S. has worked with the PRC since the 1980s to increase opportunity and economic standing for all of China, once the PRC joined the WTO they began taking advantage of the U.S.'s and the global community's good intentions. Since 2001, despite U.S.'s inclusive actions over the previous two decades, the PRC has taken aggressive strides towards the U.S. to establish itself as a regional hegemony and a force with which to be reckoned. While potentially surprising to the west, these actions are in line with Chinese self-perceived cultural identity, outlook, and regional objectives across history.

From the establishment of the PRC until around the year 2000, the PRC rebuilt Chinese diplomatic and economic capabilities through a nationalistic perspective, which can easily be equated to a Han cultural identity. This fits with the previously discussed historical cycle. The next logical steps are to build informational and military capabilities to enable power projection and the reestablishment of Chinese Great Power status and regional hegemony. This is the precursor in the historical cycle to establishing a Han culturally-led New Golden Age in China and abroad.

U.S./PRC Relations 2000-Present —The Current Situation

Once the PRC rose to more international prestige, became more integrated into the western economic systems by joining the WTO, and was granted permanent trade status by U.S. president George W. Bush, they began to immediately leverage their influence in the international community (Lu, n.d.). In 2001, a U.S. military EP-3 reconnaissance aircraft collided with a Chinese interceptor aircraft, resulting in the death of the Chinese pilot and the forced landing of the U.S. aircraft in China (Council on Foreign Relations, n.d.). Given the death of the Chinese pilot and the sensitive nature of U.S. electronic equipment aboard the aircraft, a tense standoff occurred where 24 U.S. service members were detained by Chinese authorities (Monroe 2014). The U.S. service members were eventually released after 12 days of tense negotiations, but the PRC used this situation to increase their leverage over the U.S. and their standing in the international community.

In 2005-2006, the U.S. and the international community turned to the PRC as a regional moderating power and mediator between North Korea and the rest of the world, during nuclear talks and Pyongyang's first nuclear tests (Council on Foreign Relations, n.d.). This recognition validated the PRC's self-perception as a destined regional hegemony and further emboldened the PRC to take the next logical steps in the cycle of reestablishing China's regional dominance. In 2007, the PRC increased their military spending by 18 percent, allocating over $45 billion to increasing military build-up, capability, and capacity; all the while stating goals of peace, national security, and regional integrity (Council on Foreign Relations, n.d.). This thinly veiled excuse for military build-up portends aggressive expansionism, in line with historical cycles under Han cultural identities. In 2008, the PRC continued to gain leverage over the U.S. by becoming the largest U.S. creditor in the world—holding approximately $600 Billion in American debt (Council on Foreign Relations, n.d.).

The PRC continued to rise and became the world's second-largest economy with a 2010 GDP of $5.88 trillion (Council on Foreign Relations, n.d.). This gained the attention of President Obama's administration, to the point that Secretary of State Hillary Clinton outlined a U.S. "pivot to the Pacific" to counter the PRC's explosive expansionism, resulting in plans for the Trans-Pacific Partnership (TPP) trade agreements between the U.S. and eight other nations (Monroe 2014). Tension in U.S./PRC trade relations continued to rise based on increasing deficits favoring the PRC and currency manipulations driven by the PRC (Council on Foreign Relations, n.d.). In late 2012, Xi Jinping assumed the duties of President of the PRC, among other titles, and began diplomatic conversations with the Obama Administration while apparently allowing his government to simultaneously conduct malicious cyber operations against the U.S., even after agreeing to cease such operations (Council on Foreign Relations, n.d.). Concurrently, the PRC's holding of U.S. debt increased to $1.3 trillion in 2013, demonstrating the PRC continued to maneuver against the U.S. by their leveraging of economic and informational IOP (Monroe 2014).

In 2014, the U.S. indicted five PLA-associated Chinese hackers allegedly behind technology theft from U.S. companies and major data theft from U.S. government offices (Council on Foreign Relations, n.d.). Throughout this period, the PRC continued their military build-up and territorial claims in the South China Sea through the construction of artificial islands to extend territorial claims into formerly open waters—attempting to extend their hegemonic control into areas previously non-contested (Council on Foreign Relations, n.d.). While the PRC claimed civilian purposes, U.S. surveillance determined otherwise and the U.S. openly warned the PRC against further militarization of these international spaces (Council on Foreign Relations, n.d.).

When President Trump was elected in 2016, he immediately took a more aggressive approach to limit Chinese expansionism, while attempting to maintain open diplomatic dialogue. The Trump administration affirmed the PRC's "One China" policy and hosted President Xi numerous times, but imposed tariffs against the PRC to mitigate the U.S. trade deficit (Council on Foreign Relations, n.d.). President Trump's 2017 NSS specifically states that the U.S. is in "competition" with the PRC, vice previous administrations' leaning toward cooperation (Trump 2017, 2). Tensions rose between the U.S. and the PRC because of trade wars, rhetoric, and Chinese governmental maneuvering via Chinese telecommunications company Huawei and other PRC controlled companies (Council on Foreign Relations, n.d.). However, a preliminary trade deal was signed between the U.S. and the PRC in early 2020, in which the U.S. offered to lessened tariffs and the PRC promised to more strictly enforce intellectual property regulations; however, events soon to follow would overcome the preliminary agreements leaving no official trade agreement solidified, intellectual property issues still at hand, and tensions rising between two Great Powers (Council on Foreign Relations, n.d.).

Tensions rapidly re-escalated with President Trump attributing responsibility for the emergence and the global spread of COVID-19 to the PRC (Council on Foreign Relations, n.d.). The PRC responded by expelling U.S. journalists in mid-2020, shortly after which a U.S. Executive Order was issued ending Hong Kong's preferred trade status with the U.S., and the U.S. ordered the PRC to close its Huston, Texas, consulate on claims of espionage (Council on Foreign Relations, n.d.). Negative relations culminated in July 2020 when Secretary of State Pompeo claimed that in a speech that the U.S. "era of engagement with the Chinese Communist Party is over" due to their unfair actions in trade, intellectual property theft, human rights violations, and "aggressive moves in the East and South China Seas" (Council on Foreign Relations, n.d.).

On President Trump's last day in office, Secretary Pompeo declared that the PRC's actions against the Uyghur Chinese Muslim ethnic group constituted genocide, the same term President Biden used to describe the PRC's actions towards the Uyghurs during his 2020 presidential campaign (Council on Foreign Relations, n.d.). If President Biden continues to make claims of genocide against the PRC, then U.S./PRC relations and China's global standing could take a steep and rapid downturn. Currently, under the Biden Administration, *Interim National Security Guidance* outlines the U.S./PRC strategic competition dynamic, but a specific policy has yet to emerge at the time of the writing of this paper (Biden 2021, 10).

While the western world may be frustrated or baffled at the PRC's actions, they should not be. From a Han Chinese cultural perspective, once the invading Qing Dynasty fell in the early 1900s, and internal power struggles settled with the PRC emerging victorious, they began the reconstruction/rebuilding portion of their historic geopolitical cycle. From 1949 to approximately 2000, the PRC worked tirelessly to establish diplomatic and economic strength and continue to do so through power projection and infrastructure domestically and abroad. Once that was achieved, they turned their attention to military expansion and information dominance. This is demonstrated by their over-the-top increases in military spending, building, and acquisitions as well as their increase in malicious cyber-operations, hacking, and espionage. For the last 70 years, the PRC has systematically rebuilt their entire DIME IOP infrastructure and capabilities to regain their self-perceived historic destiny as a Great Power through CCP-driven unrestricted warfare approach to establish a CCP-based state-led civilization. These actions are in line with historically demonstrated expansionism, resistance to foreign influence, and a desire to regain historically held territories and power in pursuit of reestablishing a Chinese Golden Age. This is rooted in Han cultural identity as a self-perceived and destined birthright. To effectively counter Chinese expansionism within INDOPACOM, the U.S. must understand this reality and align their IOP at the regional and national level to deter continued actions. This begins with U.S. military and diplomatic alignment within INDOPACOM.

Analysis

To establish a baseline understanding for discussion of joint-interagency structures by which to align IOP, the following analysis focuses on three major areas. The first portion focuses on the delta in eastern and western strategic approaches to war. The second offers illustrative historical examples of U.S. strategic success when military and diplomatic efforts have been aligned and failures when they are not aligned. The third section uses facts and assumptions of the current state of U.S./PRC relations to further demonstrate the need for aligned IOP to counter Chinese expansionism.

Eastern and Western Perspectives

U.S. diplomatic and military leaders must understand the differences between the PRC perspective and the U.S. perspective. The PRC's perspective is largely informed via a hybrid strategic-philosophical approach, with focus on furthering a CCP-led civilization. Much of this approach is captured in *The Art of War*, a compilation of strategic-philosophical writings attributed to Sun Tzu, but likely pulled from many ancient Chinese warrior-philosophers. The western approach to war is largely based on the writings of 18th century Prussian General Carl Von Clausewitz and his seminal work *On War*. When compared, similarities certainly exist but sharp contrasts emerge in overall approaches to strategy and warfare writ large. While both *The Art of War* and *On War* touch on operational and tactical level discussions, the most glaring difference in their approaches is their differentiating concepts of the aim of war. The eastern perspective advocates that "In war, then, let your great objective be victory, not lengthy campaigns" (Sun Tzu 2000, 7). Juxtaposing that to Clausewitz's assertion that "the aim of all War is to disarm the enemy" illustrates the differences in scope and perspective of each side's driving philosophy (Clausewitz 2019, 26).

While for millennia Chinese strategists have aimed for total victory above all, since the 18th century the west has simply tried to use violence to disarm opponents to gain acquiescence to the victors' political will. From these perspectives, the direct correlation between the eastern perspective focusing on overall strategy and the western perspective focusing on military operations and tactical-level victories becomes clear. To the Chinese and many eastern cultures, the aim of war is total victory over an opponent (Sun Tzu 2000, 7). This drives them to plan with a long-term strategic perspective and leverage every means and method at their disposal to achieve absolute victory, not stove-pipe political and military functions. Using Clausewitz as a baseline for planning, the U.S. limits itself largely to hard-power military options focused on operational-level maneuver to achieve tactical-level victories. This harkens back to Clausewitzian aims of disarming adversaries to gain limited political acquiescence and may win battles, but not wars.

Ironically, since per *The Art of War* "all warfare is based on deception," once the U.S. or west writ large think they have gained political acquiescence from eastern powers, they are surprised to find that their adversaries are not abiding by the agreed-upon terms (Sun Tzu 2000, 3). The west becomes upset because they have been deceived. This very dynamic has been evidenced time and time again, most recently post 2020 U.S./PRC preliminary trade agreement where the PRC was accused of failing to uphold agreed-upon intellectual property standards. While the west does not consider war to exist unless sanctioned violence has commenced or political bodies have declared war, the eastern perspective considers war to be perpetual competition within a political continuum. Therefore, deception is perfectly acceptable in an eastern consideration of politics or open war, since they are the same thing. The eastern perspective does not limit war to the notion of a "duel on an extensive scale," as Clausewitz does (Clausewitz 2019, 26).

The all-in unrestricted eastern approach to warfare and even GPC, results in more focused and influential actions, which the U.S. can achieve if they more effectively align military and diplomatic efforts in INDOPACOM. Historically, when the U.S. has adopted a total victory approach, such as World War II, and effectively aligned military and diplomatic IOP they have succeeded. When the U.S. has not aligned these IOP, such as in Vietnam or Afghanistan, long-drawn-out conflicts and little to no achievement of strategic objectives have resulted.

Illustrative Examples

World War II is an example of success due to the effective alignment of IOP from the top down. During World War II, the U.S. government took not only a WOG approach to warfare but a Whole of Nation approach. The war had clear and measurable objectives—to win. The definition of winning was straightforward—unconditional surrender. Strategy leveraged an Allied approach to defeat existential threats to the free world. The U.S. government-aligned all IOP to defeat the Axis powers. Diplomatic actions and alliances were made clear and unambiguous. The U.S. Government and news agencies partnered to inform the American people in ways that inspired and garnered support for the U.S. and Allied troops at home and overseas. Information was leveraged against the Axis powers to deceive them at the operational level and degrade resolve. A national draft supported the military machine. The military itself focused troop training on necessary and effective skillsets to defeat the Axis powers. The economy shifted as private companies partnered, voluntarily or due to government mandate, with the government and military to shift civilian manufacturing centers to military production hubs. When military personnel shipped overseas, citizens previously not in the workforce joined the workforce to takeover necessary wartime industry jobs. Alignment across all IOP via a WOG approach garnered the support of the American people, caused active nationwide backing of war efforts, and resulted in a rapid, decisive, and successful victory in both European and Pacific theaters.

The Vietnam Conflict is an example of failure due in part to and certainly magnified by the non-alignment of IOP. During the Vietnam conflict, the war was largely driven by faulty assumptions which caused a broken and dissociative approach that lacked holistic strategy and did not align IOP (Herring 1981). President Johnson's fear of open involvement of Soviet Russian or Chinese Communists kept him from taking a committed approach to defeating North Vietnam (Herring 1981).

This approach resulted in a policy of containment which drove diplomatic efforts that did not align with military campaigns on the ground. When operational and tactical military success resulted in gaps that could be exploited by the U.S. Government across the DIME spectrum, little was done to exploit those gaps due to the dis-alignment of agencies responsible for IOP implementation. The U.S. Government's and military's general lack of understanding of Vietnamese culture and their adversary's prime motivators resulted in little to no impact across information campaigns during the conflict (Herring 1981). From an economic perspective, the U.S. claimed to support South Vietnam but destroyed their economic capability and social fabric, further contradicting and discrediting any positive information messaging that may have been preached (Herring 1981). Inside the U.S., the war saw some sustained support across the American population even in the face of televised protests and dissent (Herring 1981). However, the U.S. Government did not effectively partner with the media as it had in World War II to communicate messages that would instill confidence and garner support across the American population. While the military units may have been successful in battle, the U.S. lost the war both at home and abroad due to a lack of consistent, effective, and strategically aligned IOP.

By briefly examining World War II and Vietnam, the importance of IOP alignment is strikingly clear. Even though this discussion is just a brief snapshot of complex and dynamic situations, these illustrative examples demonstrate how historic successes have been driven by strategically aligned IOP and catastrophic failures have resulted from dis-alignment. The U.S. must heed these lessons as it presses into GPC with the PRC in the INDOPACOM region.

Analysis of the Current Situation

The following facts and assumptions are based on that assessment and establish a level baseline from which the U.S. can move forward for strategic alignment of military and diplomatic IOP to deter Chinese expansionism within INDOPACOM.

The following statements are facts concerning U.S./PRC relations. First, the U.S. and the PRC are currently in strategic GPC with each other. This fact is specially stated in both President Trump's *2017 NSS* and President Biden's *Interim National Security Strategic Guidance* (Trump 2017, 27 and Biden 2021, 10). Second, to succeed in GPC a nation must be able to effectively operate and influence below the threshold of hostility and across a wide variety of interactions but maintain the military capability to effectively defend their interests and enforce their stances (Trump 2017, 3). Third, to effectively compete in such a manner a WOG approach is ideal, in which strategically aligned IOP are necessary for success (Austin 2021, 1). Fourth, a gap currently exists between strategic alignments of U.S. military and diplomatic IOP within INDOPACOM. This fact has been effectively demonstrated by analysis of multiple source documents in the literature review.

Planning assumptions include the following. First, the U.S. and the PRC will remain in GPC. Second, the PRC will continue military growth and territorial expansionism within INDOPACOM (Cordesman, Burke and Molot 2019, 54). Third, the U.S. will continue to seek to deter PRC expansionism in INDOPACOM. Fourth, the U.S. Government recognizes the importance of strategic IOP alignment to that end and seeks effective solutions to do so.

Analysis of these facts and assumptions drives the conclusion that the U.S. Government must take action now to effectively deter Chinese expansionism within INDOPACOM. This requires strategic alignment initially of military and diplomatic IOP in INDOPACOM, but it will eventually require alignment across the DIME spectrum. To do so effectively, personnel and resources must be dedicated to these ends. This can be done in one of two ways, both of which require joint-interagency approaches.

Discussion of Options

This paper provides two options. The first option combines the existing regional structure within the DOS and aligns that combined structure to the DOD-led INDOPACOM GCC to facilitate diplomatic and military IOP alignment. A second option leverages a proven structure to establish a Joint-Interagency Task Force (JIATF) with a singular GPC focus to holistically align efforts within INDOPACOM.

Option 1 – Combine & Align

The Combine & Align approach directs the combining of the DOS's Bureau of South Asia Affairs and Bureau of East Asian and Pacific Affairs into one new Bureau of Asian and Pacific Affairs. The new combined Bureau is then aligned to the INDOPACOM GCC. The reason for this merger of the two DOS bureaus into one is to create a singular DOS entity in the INDOPACOM region to coordinate, synchronize, and align strategic planning and follow-on actions, across the DOS and DOD at every level within INDOPACOM.

This first option meets the requirement to facilitate military and diplomatic IOP alignment within INDOPACOM. It also ensures an economy of force option by streamlining two DOS bureaus into one. However, combining both DOS bureaus may limit diplomatic bandwidth in very dynamic regions, and meet with significant pushback from DOS leadership due to a potentially perceived diplomatic draw-down. Also, combining two DOS bureaus and aligning them to a GCC may give the impression of forcing diplomatic actions into subordination to military actions within GPC. This is not the desired perception. When considering future application of the "combine and align" option across other regions, it may not be easily duplicated in other theaters because of differences and overlaps across other GCCs and DOS regions, as well as specific and informational and economic considerations across other regions. Therefore, the "combine and align" option holds limited repeatability to as an organizational construct across a global enterprise.

Option 2 – A New Joint Interagency Task Force West with a GPC Focus

A second option establishes a new Joint-Interagency Task Force West within INDOPACOM focused on GPC (JIATF-W II GPC) with the PRC which is specifically tasked with aligning diplomatic and military efforts to deter PRC expansionism within INDOPACOM. This new GPC focused JIATF-W, JIATF-W II GPC will be separate and distinct from the current JIATF-W, which focuses on counter-drug operations; however, the two can plan and operate in mutually supporting manners.

The JIATF-W II GPC option meets the requirement to facilitate military and diplomatic IOP alignment within INDOPACOM. While not as economical in personnel staffing numbers, due to an increased requirement for representatives across multiple Departments, it contains many benefits above and beyond the first option discussed. The proposed JIATF structure allows DOS and DOD to maintain or increase current personnel numbers and therefore is more likely to be an acceptable option to each department. Second, it utilizes a JIAFT structure which has proven to be very effective when aligning IOP across multiple government agencies and leveraging a WOG approach (Munsing and Lamb 2011, 1). Additionally, the JIATF structure does not necessarily give the impression of placing one department as subordinate to another, which is desirable for interagency collaboration. A GPC-focused JIATF structure within INDOPACOM allows for flexibility and ready integration of other IOP-focused departments (information and economic) once diplomatic and military IOP are aligned. The JIATF structure also enables the integration of allies and partners in a Combined JIATF (CJIATF) framework. This is key since a partnered approach is essential in deterring Chinese expansionism within INDOPACOM (Biden 2021, 8).

The only potential drawback to the JIATF-W II GPC option is the required increase in personnel. While this may be a detractor to some, it will likely gain support from DOS and DOD leadership since it gives a viable and tangible reason for personnel increases tied directly to national-level strategic directives. The establishment of JIATF-W II GPC enables effective talent management of personnel within DOD by leveraging the expertise of personnel with experience and training in foreign and regional affairs. Many U.S. military services train officers and enlisted personnel in certain regionally specific programs. However, this expertise is not always effectively leveraged due to competing requirements and legacy perspectives for military career advancement. This JIATF-W II GPC structure offers a venue for useful and real-time utilization of military-trained, regionally specific subject matter experts.

The JIATF-W II GPC structure can also be replicated and tailored across other GCCs to effectively and strategically align U.S. IOP implementation in multiple global regions. This idea will be discussed more in-depth in Future Prospects.

Comparison of Options

Based on the discussion of the two options outlined above, the option comparison chart below assists in determining the most acceptable for decision-makers. Evaluation Criteria are listed, and options are assessed to either meet the criteria or not meet the criteria. Meeting one criterion warrants a single mark. Marks are tabulated and options are scored accordingly. Across the nine criteria listed in the table below, Combine & Align (Option 1) received a score of two; and JIATF-W II GPC (Option 2) received a score of seven. Based on these criteria and this decision-matrix tool, JIATF-W II GPC (Option 2) emerges as the preferred option for facilitating military and diplomatic IOP alignment within INDOPACOM.

Evaluation Criteria	Option 1- Combine & Align	Option 2- JIATF-W II GPC
Facilitates effective alignment of military and diplomatic IOP to deter Chinese expansionism in INDOPACOM within GPC construct	X	X
Provides a flexible structure to integrate other IOP when ready		X
Provides a flexible structure to integrate other allies and partnered nations if applicable		X
Is a proven joint-interagency structure that facilitates a WOG approach		X
Maximizes economy of personnel (requires less personnel)	X	
Maximizes talent of personnel (the right person for the job)		X
Likely to be accepted by DOS and DOD leadership		X
Easily replicated and tailored across other regions		X
Total	2	7

Figure 1

Recommendations for Implementation

Based on the analysis outlined above, this paper recommends that the Executive Branch of the U.S. Government direct the establishment of a JIATF-W II GPC within INDOPACOM. It is recommended that the JIATF-W II GPC be specifically tasked and staffed to focus on GPC with the PRC by effectively aligning diplomatic and military IOP to deter Chinese expansionism within INDOPACOM. It is recommended that DOS and DOD personnel assigned to staff these positions be specially screened and evaluated to ensure the right person is selected as team members and leaders of JIATF-W II GPC. In-depth personnel selection criteria are beyond the scope of this paper; however, international relations study and training, Asian-Pacific regional/cultural specialization, understanding of DIME and IOP implementation writ large, and understanding of strategic and operational level Joint Warfare should be highly sought-after qualities across personnel selected for duty with JIATF-W II GPC. It is recommended that JIATF-W II GPC be established, staffed, and fully operational as soon as possible to effectively deter aggressive PRC actions occurring within INDOPACOM.

This can be implemented in multi-phased approach. Initially, the JIATF-W II GPC must be established with members of the DOD and DOS to align U.S. military and diplomatic efforts within INDOPACOM. Next, the addition of informational and economic experts to the JIATF-W II GPC should occur to facilitate alignment across all IOP within INDOPACOM. This will bring an aligned WOG approach where all IOP are leveraged to deter and counter the PRC's expansionist efforts. Finally, once U.S. efforts are aligned and internal process established, representatives from U.S. regional partners and allies may be added to the JIATF-W II GPC resulting in a Combined JIATF-W II GPC (CJIATF-W II GPC). This demonstrates U.S. leadership and international consensus in countering the PRC's efforts, as well as re-enforces international norms and the standing international order which the PRC is trying to challenge.

Potentially Contrasting Views

Some may oppose the JIATF-W II GPC recommendation or suggest that bureaucratic government entities are too self-focused or that policy and funding lines are too rigid for this construct to work. However, these are not sufficient reasons to excuse failure to adapt in dynamic competitive environments or cede free-trade sea-lanes and international influence to the PRC. Successful alignment precedence exists within the JIAFT construct. Specifically, in JIATF-South where exponential gains in countering illegal narcotics and other trafficking have been achieved by taking a unified approach and establishing comprehensive performance variables across the organization, team, and individual (Munsing and Lamb 2011, 32-35). Additionally, if U.S. leaders, thinkers, and decision-makers desire to gain and maintain the advantage in GPC against the PRC's current comprehensive implementation of IOP, they must be willing to overcome organizational bias and adopt a collaborative perspective to synchronize and align military and diplomatic efforts for maximum effect in the INDOPACOM region (Locher 2010, 31, 44). The JIAFT-W II GPC approach can be established and refined in INDOPACOM, but then implemented across multiple

GCCs and tailored for the unique requirements of each global region.

Future Prospects

Should the status quo remain, and the demonstrated gap across the strategic and operational alignment of military, diplomatic, and other IOP within INDOPACOM be allowed to persist, the PRC's aggressive expansionism will continue, and the U.S. will rapidly lose regional influence, security, allies and partners, and economic opportunity in the region (Henwood 2015, 4). This could rapidly foment a global geopolitical precedent resulting in decreased U.S. influence regionally and globally. Just as China once lost its regional influence with the rise of western powers in the east, if left unchecked the PRC could now turn the tables and limit U.S. influence across INDOPACOM. However, if U.S. policy and decision-makers heed the analysis and recommendations of this paper, the recommended JIATF construct can be used to effectively align military and diplomatic efforts within INDOPACOM to deter PRC expansionism. The recommended JIATF construct provides flexibility for inclusion of partners and allies in an expanded CJIATF formation, should that be desired. It also provides an exportable and tailorable organizational model for use in other theaters across the world, within GPC and beyond.

The new INDOPACOM JIATF-W II GPC can be used as a template for other GPC centric IOP alignment efforts globally. Based on lessons learned, the JIAFT-W II GPC construct can be expanded to multiple GCCs and specifically tailored for regional needs. As the concept evolves, eventually information (via DOS and DOD) and economic (via Department of Commerce) representatives can be added to the teams to effectively align all IOP across all GCCs. This will enable alignment across Diplomatic, Information, Military, and Economic IOP within all GCCs to meet regional and corresponding national strategic objectives.

As the JIATF GPC construct begins spreading across multiple regions and GCCs, the initiative will require a centralized joint-interagency management office at the national level to resource, equip, and direct forward stationed JIAFTs within different GCC regions. This will ensure that organizational infrastructure exists and is postured to manage emergent GPC focused JIATFs across multiple GCCs as they are established. This national office can be specifically tasked to oversee the achievement of cross-functional national and regional objectives which require multiple departments to synchronize and align strategic implementation of U.S. IOP. This national-level office will focus on aligning all aspects of IOP, which should be extensions of diplomatic policy, via forward-deployed JIATFs. This concept may appear to impact organizational hierarchy, but it does not have to. If implemented properly, this national-level office would benefit all Departments by serving as a Joint Interagency integrator and clearing-house to align and synchronize multiple Departments' initiatives toward overarching national strategic end states. This office would serve as a national-level strategic integrator and shape strategies and plans to align the implementation of all U.S. IOP towards the achievement of national strategic goals. The forward deployed JIATFs would serve as the expeditionary arm of this office, but still be aligned and forward deployed with the appropriate GCC and DoS Region. This construct results in multi-aspect reporting chains, but similar reporting constructs already exist at many echelons of military and diplomatic operations.

An appropriate name for this centralized national office is the Joint Interagency Expeditionary Diplomatic Initiative (JIEDI). This office can serve as the baseline for building, managing and implementing a new type of governmental operative essential in GPC. These personnel must be grounded in, but not limited by, foundations already trained within the DOD. This hybrid strategic thinker-actor must be well versed in government and politics, region and culture, information warfare, military action, and economic systems. This type of strategic influencer must understand how to use IOP to inspire and leverage adversaries and allies alike to achieve U.S. objectives. This new JIEDI office can establish and operationalize a new breed of American soldier-statesman—the warrior-diplomat—within and from its forward stationed JIATFs. This human-based capability will provide the U.S. a relevant and lasting advantage across multiple domains in and beyond Great Power Competition.

Conclusion

The rise of the PRC, and its efforts to establish itself as a regional and global force have thrust it and the U.S. into a GPC dynamic. The PRC has leveraged historic Han-Chinese identity, as many Chinese leaders have across history, to provide a driving force for cultural unification of their populations to expansionist ends. Since these expansionist ends currently challenge the established the standing rules-based international order, threaten U.S. interests, harass U.S. partners and allies, and limit global economic opportunity, the U.S. has a duty and obligation to counter and deter the PRC's expansionist efforts.

This paper has reiterated the gap in U.S. IOP alignment, specifically across military and diplomatic efforts, as well as the necessity for strategic and operational alignment of these efforts. Chinese historical and cultural context has provided a deeper understanding of root-level Chinese cultural prime drivers and motivations to U.S. decision-makers. Examples of U.S. success and failures based on IOP alignment or dis-alignment have illustrated the paramount importance of holistically aligned efforts. Analysis has provided a baseline from which to consider joint-interagency options to align IOP. Joint-interagency options have been discussed and compared, with the JIATF-W II GPC emerging as a logical and proven structure to align U.S. military and diplomatic efforts to deter Chinese expansionism within INDOPACOM. The JIATF-W II GPC not only allows the U.S. to "put our own house in order," but when the organization is ready, the construct allows for upgrading to a Combined approach with U.S. allies and partners via growth into a CJIATF (Bader 2018, 6). The recommendations in this paper provide Whole-of-Government vehicles for the U.S. to effectively align military and diplomatic efforts to deter Chinese expansionism within INDOPACOM.

Furthermore, the analysis and assessments in this document have resulted in a repeatable and regionally tailorable JIAFT GPC structure, which can be implemented across any GCC to effectively align all IOP in pursuit of regional and national strategic objectives. The benefit of a national-level collaborative JIEDI program to resource and oversee the network of multiple forward JIATFs focused on IOP alignment has been outlined and advocated. If established and properly resourced, the capability of the JIEDI program to identify, grow, equip, and employ teams of new highly skilled soldier-statesmen—warrior-diplomats—will give the U.S. a sustained, repeatable, and adaptable advantage within and beyond Great Power Competition. If implemented properly, the structural vehicles proposed in this paper will enable the U.S. to coordinate, synchronize, and align a strategic Whole-of-Government approach and operationalize highly trained warrior-diplomats to implement this strategy. This will enable the achievement of regional and national strategic objectives resulting in security and opportunity, peace and prosperity for the people of the United States and those of its partners and allies.

Bibliography

Austin, Lloyd J. III. "Memorandum for All Department of Defense Employees." Washington, D.C.: Office of the Secretary of Defense, 4 March 2021.

Bader, Jeffery. "U.S. – China Relations: Is It Time to End the Engagement?" Brookings Institution, 2018.

Biden, Joseph R. "Interim National Security Strategic Guidance." Washington, D.C.: The White House, March 2021.

Chief of Mission, U.S. State Department. "Integrated Country Strategy – China." Washington, D.C.: Department of State, approved 2018/updated 2020.

Clausewitz, Carl von. *On War*. Edited and translated by J.J. Graham. Online: Global Grey ebooks, 2019, https://www.globalgreyebooks.com/on-war-ebook.html

Congressional Research Services. "Renewed Great Power Competition: Implications for Defense" – Issues for Congress, R43838 (29 October 2020).

Cordesman, Anthony H., Arleigh A. Burke, and Max Molot. "China and the U.S.: The U.S. Department of Defense, Defense Intelligence Agency, and INDOPACOM Command View of China's National Security Strategy." Center for Strategic and International Studies, 2019.

Council on Foreign Relations. "Timeline – U.S. Relations With China 1949-2020." Washington, D.C.: Council on Foreign Relations (accessed 17 December 2020), https://www.cfr.org/timeline/us-relations-china

Encyclopedia Britannica online. "An Lushan Rebellion summary." (accessed 8 March 2022), https://www.britanica.com/summary/An-Lushan

Encyclopedia Britannica online. "The Song (960–1279), Liao (907–1125), and Jin (1115 1234) dynasties." (accessed 8 March 2022), https://www.britanica.com/art/Chinese-architecture/The-Song-960-1279-Liao-907-1125-and-Jin-1115-1234-dynasties

Encyclopedia Britannica online. "United Front; Chinese history [1937-1945]." (accessed 7 March 2022), https://www.britanica.com/topic/UnitedFront-Chinese-history-1937-1935

Gronewold, Sue. "The Ming Voyages." Asia for Educators. Columbia University (accessed 31 March 2021), http://afe.easia.columbia.edu/special/china_1000ce_mingvoyages.htm

Henwood, B.P. "The Dragon Challenger: China's Long Game Toward Global Power." Canadian Forces College, 2015, https://www.cfc.forces.gc.ca/259/290/317/305/henwood.pdf

Herring, George C. "The 'Vietnam Syndrome' and American Foreign Policy." *Virginia Quarterly Review, Autumn 1981, Volume 57, #4.* Charlottesville, VA: University of Virginia Press, 2003, https://www.vqronline.org/essay/nglish-syndrome-and-american-foreign-policy

Histroy.com Editors. "China: Timeline." History.com. A&E Television Networks, 22 March 2019, https://www.history.com/topics/china/china-timeline

Histroy.com Editors. "Tiananmen Square Protests." History.com. A&E Television Networks, updated 9 Jun 2020, https://www.history.com/topics/china/nglishn-square

Jones, Bruce. "China and the Return of Great Power Competition." Washington, D.C.: The Brookings Institute, February 2020, https://www.brookings.edu/wpcontent/uploads/2020/02/FP_202002_china_power_competition_jones.pdf

Lu Qichang, editor. "Chronology of China-US Relations." Translated by Zheng Guihong for china.org.cn (accessed 14 April 2021), www.china.org.cn/nglish/china-us/26890.htm

Mattis, James. "Summary of the 2018 National Defense of Strategy of The United States of America: Sharpening the American Military's Competitive Edge." Washington D.C.: The Office of the Secretary of Defense, 2018.

Monroe, Erin. "U.S.-China Relations: A Brief Historical Perspective." U.S.-China Policy Foundation, accessed 14 April 2021, https://uscpf.org/v3/wp-content/uploads/2014/08/backgrounder-on-US-China-relations

Munsing, Evan and Christopher J. Lamb. "Joint Interagency Task Force – South: The Best Known, Least Understood Interagency Success," Strategic Perspectives 5, Institute for National Strategic Studies. Washington, D.C.: National Defense University Press, 2011.

Office of the Historian. "Chronology of U.S.- China Relations, 1784-2000." Foreign Service Institute, Washington D.C.: Department of State (accessed 17 December 2020), https://history.state.gov/countries/issues/china-us-relations

Plante, Trevor K. "U.S. Marines in the Boxer Rebellion." *Prologue Magazine Winter 1999, Vol. 31, No. 4.* Washington, D.C.: The National Archives, 1999, https://www.archives.gov/publications/prologue/1999/winter/boxer-rebellion-1.html#:~:text=The%20total%20number%20of%20marines,officers%20and%201%2C151%20enlisted%20men

The Joint Staff. "Description of the National Military Strategy 2018." Washington, D.C.: The Pentagon, 2018.

The Joint Staff. *Joint Publication 3-08 Interagency Cooperation*. Washington, D.C.: The Pentagon, 2016/validated 2017.

The Joint Staff. "Joint Doctrine Note 1-18 Strategy." Washington, D.C.: The Pentagon, 2018.

Tillerson, Rex W. "Joint Strategic Plan FY 2018-2022." U.S. State Department, U.S. Agency for International Development. Washington, D.C.: Office of the Secretary of State, 2018.

Trump, Donald J. "National Security Strategy of the United States of America." Washington, D.C.: The White House, December 2017.

Tzu, Sun. *(On) The Art of War* – Translated from the Chinese by Lionel Giles (1910). Leicester, England: Allandale Online, 2000, https://electricscotland.com/history/scotreg/TheArtOfWar.pdf

---

Jonathan E. Schillo is a United States Marine. He holds a Master of Arts in International Relations, a Master of Military Operational Art and Science – focus in Joint Warfare, and a Bachelor of Science in History. He is a General Robert H. Barrow Fellow (AY22) with the Marine Corps University's Krulak Center for Innovation & Future Warfare. His primary areas of research include the application of international relations principles, contextualized through an historical perspective, to develop strategies and operational frameworks to gain and maintain advantage in dynamic environments. He welcomes opportunities for continued research and collaboration, jeschillo@gmail.com.

The Ideology of "Strategic Conservatism" from Russia's Imperial Perspective

Eugene Alexander Vertlieb

Translated by Dennis T. Faleris

> "Unfortunate is the country in which any conservatism has become cruel and abusive ... If so, there is a *revolution* being prepared in that country." (Philosopher N. Berdyaev)

> "Conservatism without an aggressive imperial idea turns into a preaching of petty-bourgeois conformism based on the logic '*if only it doesn't get worse*.'" (Historian A. Minakov)

> "Dmitry Medvedev is no less—in a good sense of the word—a Russian nationalist than I am." (V. Putin)

Russian Religious Foundations of the Current Conservative-Liberal Clash

1) Christian religious denominations—Orthodoxy, Catholicism and Protestantism—originated from a single spiritual basis, a proto-Christianity, if you will. In 1054, the final split of the Eastern and Western churches took place, with Orthodox as the Eastern, and Catholic as the Western church. This great schism was the main cause of many inter-civilizational wars.

2) In 1653, Russia annexed Left-Bank Ukraine,[1] on whose territory the Greek rite was practiced. The Greek rite was also widespread in the Balkans and the Middle East. The Russian rite differed from the Greek one. For example, the name of Christ was rendered differently—Jesus being spelled "Isus" as opposed to Jesus spelled "Iisus," and baptismal rites that were carried out using two fingers rather than three.[2] The unification of the Russian and Greek rites would unite all the Slavs—Eastern and Balkan—and

1 Left-bank Ukraine is a historic name of the part of Ukraine on the left bank of the Dnieper River comprising the modern-day oblasts of Chernihiv, Poltava, and Sumy, as well as the eastern parts of Kyiv and Cherkasy.

2 Patriarch Nikon (1605-1681) launched widespread reforms within the Orthodox church, revising church ceremonies, literature, and even such changes as the rendering of the name "Jesus" to correspond to the Greek spelling, written in ancient Greek as ὁ Ἰησοῦς—in Cyrillic letters, "Иисус"—rather than the usual Russian spelling "Исус." Another change made was in making the Sign of the Cross: Nikon's change called for the use of three vice two fingers so that

 doi: 10.18278/gsis.7.1.7

would lay the foundation for the Great Greco-Russian Empire. Consequently, Patriarch Nikon carried out church reform on behalf of Tsar Alexei Mikhailovich.[3] However, the introduction of changes in the liturgical books and in some rites aimed at unifying them with the modern Greek ones themselves caused a schism in the church, with Avvakum and the Old Believers pulling away from the Nikonians—followers of Nikon and the official Russian Orthodox church.[4] Echoes of that spiritual-based feud can be seen, in part, in the clash of the current Russian conservative and liberal ideologies, that is, modern-day Old Believers—the carriers of traditional Russian folk religiosity— vs. Western liberals.

3) Russian culture is one of spiritual integrity and one which views truth as a living ontological essence in the world. The followers of Nikon upset the fundamental foundations of the Russian worldview and mindset. The change in the rites of worship and the structural alterations in the hierarchy of the church administration disrupted, for worshipers, the organic connection between God and the Church. Before those "innovations," it was considered an undeniable truth that only the "two-fingered" (Old Believer) sign of the cross reflected the true dogmatics of the Christian Creed—the crucifixion and resurrection of Christ—as well as the two natures of Christ, the human and the divine. In the three-fingered (New Ritual or Nikonian) sign of the cross, from the standpoint of dogma, there was a distortion in the true meaning of the sign of the cross: it was as if making the sign with three fingers signified that the Trinity was crucified on the Cross! The church oath and the decrees of the Stoglavy Council of 1551[5] secured

---

the Russian Orthodox sign was in line with the Greek practice. The effect on Russian society was profound: The Old Believers were anathematized for almost two and a half centuries, and the consequences of the schism in the Russian Orthodox Church have not completely been overcome to this day. https://russiapedia.rt.com/prominent-russians/religion/patriarch-nikon/index.html

3 Nikon attracted the attention of Tsar Aleksey Mikhailovich in 1646 and soon became his confidant and spiritual advisor. He also became an important figure in the reformers, the Zealots of Piety. By 1652, Nikon was consecrated patriarch of Moscow and all Russia and was directed to institute reforms advocated by the Zealots and supported by the Tsar. See https://www.encyclopedia.com/history/encyclopedias-almanacs-transcripts-and-maps/nikon-patriarch

4 Patriarch Nikon brutally persecuted Avvakum's "Old Believer" followers who refused to accept liturgical reforms. https://www.britannica.com/biography/Avvakum-Petrovich and URL https://en.wikipedia.org/wiki/Raskol

5 The Stoglav Synod or Hundred Chapter Council [in Russian, Стоглавый собор] was a church council held in Moscow in 1551 with the participation of Tsar Ivan IV, Metropolitan Macarius, and representatives of the Boyar Duma. The Council produced a church code formatted as a record of questions by the Tsar to the clergy. It consisted of 100 chapters and recorded decisions by the Council governing church ceremonies, duties, and other matters. https://or-

the inviolability of the two-fingered practice: "If anyone does not bless with two fingers like Christ or does not imagine the sign of the cross, let him be damned."[6]

4) This "innovative postmodern act" of spiritual transformation changed the very concept of sin. The disoriented flock, uprooted from its religious norms, lost confidence in its ability to discern the "righteous" from the "sinner." At the Last Judgment, "the Lord will place the righteous at His right hand (on the right hand), and sinners on the left (on the left hand).[7] A latent fear arose among the parishioners going to the temple of the Lord that they would not please the "heretics with no grace"[8] because of the confusion that had arisen. They called Patriarch Nikon, who declared about himself "I am Russian in body, but Greek in soul," the Antichrist.

5) It took a long time for the split Orthodoxy to return to a state of unity. According to folklore- and ethnic-based beliefs and traditions, the broken parts were first spliced together, then spiritualized with living water, and only then was its soul returned. At the local councils of 1918 and 1971, the Russian Church recognized the equal salvation power of the old rites. The Edinoverie Church[9] returned the Old Believers to the Ecumenical Church, with "Edinoverie" referring to "one faith"—the Ecumenical Orthodox Church. Metropolitan Anthony (Khrapovitsky) restored the truth about Patriarch Nikon.[10]

---

thodoxwiki.org/Stoglavy_Sobor

6 In the original Russian text, the author, Dr. Vertlieb, provides the reader with this quote not in Russian, but in Church Slavonic, the liturgical language still in use today in the Russian Orthodox Church.

7 Here the author is using archaic Russian adverbs (Church Slavonic) followed by their contemporary Russian counterparts in parentheses though they are translated the same. Also, the author notes that the standing of sinners on the left hand also refers to the position of the hand on the left shoulder when making the sign of the cross.

8 Here the "heretics without grace" refer to the Nikonist reformers who were instituting changes that were unsettling to the traditional worshipers. Discord continues today among various manifestations of the Orthodox church with the same label being used to condemn the alternative churches.

9 Edinoverie [in Russian, Единоверие] is an arrangement between certain communities of Russian Old Believers and the Russian Orthodox Church, whereby the communities are treated as a part of the mainstream church, while maintaining their own traditional rites. Thus, they are often described as Old Ritualists rather than Old Believers [in Russian, староритуалисты and старообрядцы, respectively.] https://orthodoxwiki.org/Edinoverie

10 On May 3, 1910, Metropolitan Anthony (Khrapovitsky) delivered a lecture entitled "Restored Truth about Patriarch Nikon" [in Russian, "Возстановленная истина о Никоне"] which offered new, ground-breaking interpretations of Nikon's life—his words and actions. See (1910) *Mirnyj Trud* [in Russian, Мирный труд] 9, 140-171. Note: On the 75th anniversary of Anthony's death, the original text was reprinted: (9 August 2011) *Russkaya Narodnaya Liniya.*

With the advocacy of the thesis "Byzantium is the foremother of Russia" (which was helped by the convincing film of Metropolitan Tikhon *The Death of the Empire: A Byzantine Lesson*"),[11] Nikon's Christian cosmopolitanism was rehabilitated as "apostolic neo-Byzantism." Patriarch Nikon, it turns out, then also defended the idea of a "symphony of authorities" (one which was "two-headed"—spiritual and secular), now in demand in Russia, headed by an Orthodox tsar-autocrat, the one anointed by God.

6) The opposition of both confessions to Catholicism furthered the unification of the Orthodox branches of the clergy. The opposition of these antipodes is ontological. According to the philosopher-theologian V. V. Zenkovsky,[12] Catholicism, being content with only the external trappings of piety, did not accept the teachings of Christ concerning the inner conversion of a person to truth and love. Thus, it rejected the gospel of Christ about freedom whereas Orthodoxy "is truly free in Christ" and consequently, in the eyes of Orthodox authenticity, Catholicism "lost the truly Christian principle."

7) On the other hand, the idea of theocracy (God+control) appeared, which, when applied to society, gave rise to the temptation to subordinate humanity to a single authority (globalism of a "world government"). All other troubles grew out of this. According to Zenkovsky, Christianity, reformed by Catholicism, is "socialist" in spirit in the sense of forcibly leading people to a social "paradise" of material well-being. And from the unintentional consequences of the Catholic reformation of Christianity—"Catholic socialism"—it was only a stone's throw away from revolutionary socialism with a human face: "that chaos of freedom, that chaos of immorality in which modern humanity lives, was caused by the fact that Catholicism rejected Christ's teaching about freedom." Modernization within the Catholic church life—the involvement of the faithful in church-based social and religious activities—was a

https://ruskline.ru/analitika/2011/08/10/vozstanovlennaya_istina

11 The documentary film "The Death of the Empire. A Byzantine Lesson" was written and narrated by Russian-Greek Archimandrite Tikhon. According to one review, the film deals with the Byzantine Empire's "degradation and how it lost its 'ability to respond to the calls of history'.... Due to a reference to the Emperor Constantine as The Drunkard, not a few critics saw in the film a portrayal of the late President's Yeltsin's crumbling Russia and considered the documentary an attempt to help President Putin's hand-picked successor and [then] current President Dmitri Medvedev win the election." https://neomagazine.com/2008_06_june/2008_06_10.html

12 V. V. Zenkovsky is a noted historian of Russian philosophy and author of several works on the subject, including his two-volume *History of Russian Philosophy*, first published in English by Columbia University Press in 1953.

catalyst for revolutionary positivist social upheavals. In the "Legend of the Grand Inquisitor" from F. M. Dostoevsky's novel *The Brothers Karamazov*,[13] the Catholic assertion that humanity is incapable of Christian freedom is brought to the fore and explored. Having lost faith in Christ, the Inquisitor wants to solve social problems without Christ, but with the help of a "sick" Catholic church that has departed from his precepts. The meaning of this allusion is to prove the primacy of the need to "instill the ideal of beauty into souls" over the calls of socialists to "Feed, then ask for virtue!"[14]

8) Catholicism, as the Slavophile[15] A. Khomyakov[16] asserted, betrays the principle of freedom in the name of unity, while Protestantism is the opposite. In terms of this concept, only Orthodoxy remained true to the spirit of early Christianity, being a harmonious combination of unity and freedom in the principle of Christian love. Having rejected the principle of "sobornost," Catholicism was pervaded by rationalism; Protestantism did nothing more than further develop Catholic rationalism, leading the way from unity to freedom. "The West," culturologist N. A. Narochnitskaya maintains, "is freedom 'from what' (i.e., the absence of restrictions), while Russia is freedom 'for what' (i.e., why freedom is needed)."[17] Or-

13 *The Brothers Karamazov* was first published in serial form by The Russian Messenger Literary-Political Journal [in Russian, Русский Вестник-Литературный и Политический журнал] between 1879 and 1880. The complete novel was published in a separate edition in 1880. Set in 19th-century Russia, *The Brothers Karamazov* is a philosophical novel that delves into questions of God, free will, and morality.

14 This quote is taken from Doestoevsky's "The Grand Inquisitor," a poem contained within the text of his novel *The Brothers Karamazov*. "The Grand Inquisitor" is recited by the character Ivan Karamazov, who questions his brother Aleksey, a novice monk, about the possibility of a personal and benevolent God. Among other beliefs, the Grand Inquisitor defends the idea that you need only give man bread and control his conscience to rule the world.

15 Slavophilia was a 19th-century intellectual movement that considered western Europe, which had adopted the Roman Catholic and Protestant religions, as morally bankrupt. The Russian people, by contrast, adhered to the Russian Orthodox faith and thus, through their common faith and church, they were united in a "Christian community." Among its leaders were Aleksey S. Khomyakov, the brothers Konstantin S. and Ivan S. Aksakov, the brothers Ivan V. and Pyotr V. Kireyevsky, and Yury F. Samarin. https://www.britannica.com/topic/Slavophile

16 Aleksey Stepanovich Khomyakov was a 19th century Russian religious writer whose ideas revolved around the notion of "sobornost," (in Russian, соборность) or "catholicity." Khomyakov believed that sobornost called for "cooperation within the Russian Christian community or 'obshchina' [In Russian, община], united by a set of common convictions and Orthodox Christian values, as opposed to the cult of individualism in the West." https://www.definitions.net/definition/sobornost and https://www.britannica.com/topic/Slavophile

17 N. A. Narochnitskaya, a Russian historian known for her radical conservativism and support of Russian military action in the Chechen wars, and opposition to NATO action in the former Yugoslavia. In numerous interviews, Narochnitskaya has claimed "that the West wants to sub-

thodoxy is the freedom to serve Christian virtue "in the name of the Father and the Son of the Holy Spirit."

9) If so, in Orthodoxy, the whole truth of freedom having been given to man is preserved but the chaos that comes with it is overcome. Orthodoxy does not need "solitude" for some "self-absorbed development of a personal basis." The Russian Orthodox path is healing—it assumes that social contradictions are resolved not by forcibly imposing happiness on humanity (consumerism, hedonism, success, egalitarian progressivism), but by reconciling everyone and everything in the bosom of the Church. The Orthodox "sobornost" consciousness "churchifies" life in Christ. It is this positive ideal that inspired the thinker, Dostoevsky, and which he understood not as the external subordination of all life to the Church (as Catholicism represents), but as the free and internal assimilation of Christian principles in life in all its forms of being. Russian universal humanity gives rise to the fulfillment of Christ's precepts on earth. Through repentance, man has returned to the God-ethical transfigured self. The deeper a person has fallen, the greater the feat of his moral resurrection. Suffering in atonement for the sin and evil committed. If it is perceived sincerely and deeply, freely manifested and truly suffered, moral healing is possible. Such are the spiritual tablets of the Russian national worldview.

10) The Russian traditional-religious model that shapes the worldview was based on the absolute opposition of the poles of good and evil. Russian consciousness is extreme—"all or nothing." It does not like the middle-ground (where there is gray devilry), it is uncompromising in its essence. Compromise is seen as being unprincipled, lacking the ability to show will or stand on one's own beliefs. There is no "golden mean" in pragmatics, no expediency. Separating freedom from the Cross—"a sinful paradise"—is a moral collapse for the Russian consciousness which is unclouded by the revisionism of adaptation. The primordial purity of Orthodoxy is doomed to "competition with the Latins."[18] For the "believing mind" is truly God-believing and

jugate Russia, impose its rules on it, even 'dismember' it. She is often seen on state TV channels corroborating Putin's claims that the opposition movement in Russia is funded by coordinated by the NATO nations." https://www.interpretermag.com/gay-slavs-are-better-than-gay-teutons/ and https://en.wikipedia.org/wiki/Natalya_Narochnitskaya

18 "Competition with the Latins" refers to an 11th century essay directed against Catholics. Although Metropolitan of Kiev Georgiy Grek is considered to be the author of this work, many researchers consider this attribution to be incorrect. Metropolitan Georgiy, who arrived in Rus' from Byzantium in around 1062, is believed to have sat on the metropolitan throne from 1062 to 1073. https://dic-academic-ru.translate.goog/dic.nsf/ruwiki/1642334?_x_tr_sl=ru&_x_tr_tl=en&_x_tr_hl=en&_x_tr_pto=op,sc and https://dic-academic-ru.translate.goog/dic.nsf/ruwiki/12

free in the Cross. Orthodoxy recognizes itself as genuine, unreformed Christianity, while the reformation of Catholicism unwittingly gave rise to the revolutionization of the consciousness of Europeans. The mission of Orthodoxy is to protect the spirit in pristine moral purity, and the world order in Divine harmony with truth and justice.

Western/European ideologies forced upon Russia have weakened it; Russia must return to its roots

11) For the Russian "sobornost" consciousness, not only are Western fundamental and ideological innovations unacceptable, but also the "antichrist" reforms of Emperor Peter the Great: For their ruthlessness, Peter was called the "first Bolshevik." Despite all the technological benefits to the Fatherland, he inflicted irreparable harm on the Russian people: he abolished the patriarchate, subordinated the Church to the state, and "suspended" Russian holiness. Having cut a window into Europe, he broke with Russian conservative tradition, muddied its identity, and changed the national code. His issue with the "beard"[19] obliterated the conservative "long-haired" norm and caused a regression, which led to the destruction of the traditional Russian hierarchy of values. Let him then strengthen imperial sovereignty—a trait borrowed from Europe. "He wanted," as the philosopher Jean-Jacques Rousseau noted, "first to create Germans and British, when he should have started by creating Russians. He prevented his subjects from ever becoming what they could be."[20] Emperor Peter I decided to transform Russia into a Holland (even the current flag in the Russian Federation is that same alien tricolor) at the cost of turning the state "into a plaything for endless changes" which led to revolution. From the standpoint of a conservative vision, he undermined the "spirit of the people," that is, the very foundations of an autocracy, the very moral power of the state. In connection with this,

---

92347?_x_tr_sl=ru&_x_tr_tl=en&_x_tr_hl=en&_x_tr_pto=op,sc

19 Eschner, Kat. (2017, September 5). Why Peter the Great Established a Beard Tax. *Smithsonian Magazine*. In his attempts to modernize Russia, Peter the Great endeavored to get Russians to go beardless like "modern" Western Europeans. After shaving his own beard off, he declared that all men in Russia had to lose theirs. This was a massively unpopular policy with many including the Russian Orthodox church which considered the practice blasphemous. Ultimately the tsar allowed people to opt to keep their beards by imposing a beard tax. https://www.smithsonianmag.com/smart-news/why-tsar-peter-great-established-beard-tax-180964693/

20 This quote is taken from Book 8, Jean-Jacques Rousseau's 1762 treatise, "The Social Contract" [In French, "Du contrat social"]. Here Rousseau theorizes about "the best way to establish a political community in the face of the problems of commercial society.... [The work] helped inspire political reforms or revolutions in Europe...[arguing] against the idea that monarchs were divinely empowered to legislate...[and] that only the people, who are sovereign, have that all-powerful right." https://en.wikipedia.org/wiki/The_Social_Contract

historian N. M. Karamzin, in the *Note On Ancient and New Russia,* stated: "With the acquisition of human virtues, we have lost civil ones . . . We became citizens of the world, but ceased to be, in some cases, citizens of Russia. Blame Peter." [21]

12) Karamzin blames Peter for many things, and most importantly, the creation of a Europeanized ruling stratum, which, in fact, ceased to be Russian. He blames him for the gigantic, painful socio-cultural split, the gap between the top and the bottom. "His accusation of Peter for the creation of a socio-cultural abyss between the top and the bottom, was fraught with the likelihood of revolution. And this imputation was, perhaps, one of the main bonds underpinning Russian conservative ideology until 1917. And even after," historian Arkady Monaco states, Peter "denationalized the upper social layer. Made it cosmopolitan. This meant that they were of little use for solving the problems that Russians and Russia faced." [22]

13) How did Russia return to itself—to its root archaic religion? The ancient Russian fundamental, "unwavering" pantheism,[23] which considered existence as an ontologization of the moral outlook on the world, nevertheless gave way to an "innovative" path of development. In post-Decembrist Russia,[24] the "disintegration phase" (the theory of ethnogenesis of Lev Gumilyov[25]) of Russian integrity began. Under

21 Nikolay Mikhailovich Karamzin is a Russian historian and writer who, in 1803, was given the title of historiographer by Tsar Aleksandr I. In 1811, he submitted his "Note on Ancient and New Russia," which contained a biting critique of the policies of the Tsar but vindicated autocracy and serfdom. He is considered a founder of 18th/19th-century Russian imperial conservatism. https://www.encyclopedia.com/people/history/historians-european-biographies/nikolai-mikhailovich-karamzin

22 Minakov, Arkady. (2018, July 5). Voronezh Historian Arkady Monaco: The Cult of the West as a Sickness of Russian Civilization. *Four Pens-The Voronezh Independent Socio-Political Portal.* [In Russian, "Воронежский историк Аркадий Минаков: Западничество как болезнь русской цивилизации."] http://4pera.com/news/history/voronezhskiy_istorik_arkadiy_minakov_zapadnichestvo_kak_bolezn_russkoy_tsivilizatsii/

23 Panteism refers to a philosophical doctrine that identifies God with the universe or regards the universe as a manifestation of God. It can also refer to worship that admits or tolerates all gods.

24 The Decembrist uprising refers to a December 1825 revolt staged by Russian imperial army officers, influenced by European intellectual trends, who led approximately three thousand Russian soldiers in an attempt to implement a liberal political program. With the failure of the Decembrists, Russia's monarchial absolutism continued for another century. For details, see https://www.newworldencyclopedia.org/entry/Decembrist_Revolt

25 Titov, Alexander Sergeevich. (March 2005). Lev Gumilev, Ethnogenesis and Eurasianism. *University College London, School of Slavonic and Eastern European Studies.* pp. 2, 60, 122. According to Titov, "Gumilev's account of Russian history focused on a distinction between Kievan Rus and Muscovite Russia, the role of the Mongols in the formation of the Russian ethnos, and the interpretation of Russian history in terms of phases of ethnogenesis." Titov explains further that Gumilev's "ethnos" referred to "a group of people who had a stable common culture and

Pushkin, "the end came to that 'organic' Russia, whose monument was erected in Tolstoy's *War and Peace*. Then the Russian spirit was driven underground or banished to a foreign land," writes Academician of the Russian Academy of Sciences A. Panchenko.[26] Morality is truth—the leitmotif of the revivalist "characteristically Russian" prose of life—the life "of those in the village." The voice of Prince Shcherbatov[27] from the 18th century *On the Damage to Morals in Russia* sounds like a serious warning to contemporaries about a perniciousness of reform that is alien to the nation. He rightly believed that the morals of pre-Petrine Russia were healthier and more suitable for the "preservation of the people," and that it would be better to do without reforms altogether. If the churched spiritual backbone of the people and imperial sovereignty had not been strong, then Russia would not have had to be broken over the knee twice in the 20th century alone—first by the Bolshevik-Leninists, then by the anti-Bolshevik Yeltsinites. And today, Russia is intent on a complete restoration of its integrity and imperial self-identification.

Russia's geopolitical presence has waned; NATO expansion has magnified the threat

14) A permanent conflict of civilizations has become a pre-war confrontation. Russia, as the English historian A. Toynbee[28] defines its modern-day fateful moment, is resisting the implanting of someone else's "civilizational wedge." Even without Newton's law, it is clear the counteractive

could assume any social form, from a tribe to a state." Gumilev maintained that "by looking at people's lives and attitudes at a particular time in the history of an ethnos, it was possible to determine which phase of ethnogenesis was at work. Here Dr. Vertlieb is making reference to the third phase of ethnogenesis—named variously as the "disintegration" or "break down" or "crisis" phase. According to Gumilev, this phase begins with unsuccessful attempts at a reform of social institutions, followed by civil wars and behavioural splits in the superethnos. https://discovery.ucl.ac.uk/id/eprint/1446515/1/U602440.pdf

26 Panchenko's words are taken from the Foreward to Dr. Vertlieb's book (1992) От Загоскина до Шукшина: опыт непредвзятого размышления, Библиотека 'Звезды', СПг, 403 стр., ISBN 5718300453, 9785718300451. [In English, *From Zagoskin to Shukshin: An Unbiased Reflection*], Zvezdy Library, St. Petersburg, 403 Pages.

27 Prince Mikhail Mikhailovich Shcherbatov (1733-1790) was a leading ideologue and proponent of the Russian Enlightenment. Shcherbatov's essay "On the Damage of Morals in Russia" sharply criticized the policy of the government and the customs of the court environment where he served as a historiographer and publicist.

28 Sokolov, S. V. (2011, 21 December). *Conflict of Christian and Islamic Civilizations in the 21st Century.* Medina Publishing House. http://idmedina.ru/books/materials/?3708. Here Dr. Vertlieb is paraphrasing the words of English historian, author, and specialist on international affairs, Arnold J. Toynbee, words which, according to the Sokolov article, are found in Toynbee's work (1949, January 1) *Civilization on Trial.* Oxford University Press. 263 Pages. ASIN: B0007K8VCM.

force exerted by the Russians should theoretically be no weaker than the force of influence on them. However, the fact that the countries of Eastern Europe have been coopted into the North Atlantic Alliance[29] almost unhindered by the Russian Federation speaks of the passivism of Russian foreign policy and the absence of strategic parity between the antagonists. Russia's geopolitical space is rapidly shrinking, like Balzac's *Wild Ass Skin*.[30] The West has moved close to the borders of the Russian Federation. The Kremlin's "Red Line" has signaled the danger.[31]

15) As the political scientist S. Huntington rightly argues, "the fault lines between civilizations are the lines of future fronts."[32] Isn't that why, in 1945, American General George Patton drove his army without respite to meet the Russians—to prevent Marshal Georgy Zhukov from occupying all of Europe at that time? If then, the "inter-civilizational fault line" ran near the German city of Torgau on the Elbe River—now [the line runs all of] a few minutes of missile flight time even from Romania, let alone from Poland. And if Ukraine enters NATO, the West will decrease its flight time to the Russian Federation to as few as 5 to 7 "missile" minutes to reach Moscow! In order to avoid the worst-case scenario, on December 17, 2021, the Kremlin demanded *a written* guarantee that the military development of Ukraine would stop, and that NATO's entire military infrastructure would move back to 1997 positions.[33] To put it another way: it has

29 The United Nations classifies Eastern Europe as including 10 countries: Belarus, Bulgaria, Czech Republic, Hungary, Moldova, Poland, Romania, Russia, Slovakia, and Ukraine. Current Eastern European members of NATO, then, include 7 of the 10 Eastern European countries: Bulgaria, Czech Republic, Hungary, Poland, Romania, Slovakia, and Slovenia.

30 Written by French novelist and playwright Honore de Balzac in 1831, *The Wild Ass's Skin* [in French, *La Peau de chagrin*] tells the story of a young man who discovers a piece of wild ass's skin which has the magical property of granting wishes. However, the fulfillment of the wisher's desire comes at a cost: after each wish, the skin shrinks and consumes the physical energy of the wisher. https://www.goodreads.com/book/show/99742.The_Wild_Ass_s_Skin

31 Seddon, Max; Foy, Henry; Williams, Aime. (2021, December 17) "Russia publishes 'red line' security demands for NATO. US: Moscow blames alliance for 'hostile acts' as tension simmers over military build-up near Ukraine." *Financial Times.* According to the article, "Russia...published a set of stringent demands...which would end all prospect of Ukraine or any more former Soviet states joining the transatlantic alliance...." Further, the article states that according to Putin, "[the demands] are needed to insulate Russia from the threat of attack." https://www.ft.com/content/493da5ea-6ef2-42cc-8be1-c725030cf839

32 Huntington, Samuel P. (Summer 1993). The Clash of Civilizations? *Foreign Affairs* 72(3), 22-49. In the article, American political scientist argues that people's cultural and religious identities will be the primary source of conflict in the post-Cold War world and that future wars would be fought not between countries, but between cultures.

33 Meyer, Henry and Arkhipov, Ilya. (2021, December 17). Russia Demands NATO Pullback in Security Talks with U.S. *Bloomberg News*. According to the article, "Russia demanded that

proposed [that NATO] capitulate in a war that has not yet begun. In the ancient treatise *The Art of War* by the Chinese commander Sun Tzu, it is said, "Battles and the seizing of territory cannot be considered the highest skill of military operations; the highest skill is forcing the surrender of the enemy army without a fight."

16) It is unlikely that the collective West would agree to surrender. After all, "Russia is very much inferior to NATO both in terms of human and industrial resources as well as in terms of the total power of its weapons," military expert B. Yulin[34] maintains. That is, unless there is something that equalizes the offensive forces of both parties—like the hypersonic weapons and powerful nuclear arsenal that the Russian Federation possesses. The Russian Federation is not the USSR, which threatened, in case of force majeure, to use the plan of Academician A. D. Sakharov—to create a "strait named after Stalin" between Canada and Mexico.[35] For the sake of business, the Russian Federation has been filling the tanks of "independent [Ukraine]" with Russian fuel. And only when everything "became tangled up" into a Gordian knot was the Russian Federation "forced to do something" (V. Putin's words). But if it were not for the risk to the interests of the ruling oligarchy in the Russian Federation, the Kremlin would hardly have resorted to the daring rhetoric of war—a warning about "adequate military-technical measures" in case of force majeure.[36]

---

the North Atlantic Treaty Organization roll back almost a quarter-century of expansion by withdrawing forces form eastern Europe and halt further growth...." https://www.bloomberg.com/news/articles/2021-12-17/russia-demands-nato-return-to-1997-in-security-treaty-proposals

34 Abramov, Nikolai. (2021, November 12). Historian Yulin assesses the risk of a military conflict between Russia and NATO [Russian title: "Историк Юлин оценил риск военного конфликта между Россией и НАТО"]. https://www.gazeta.ru/politics/news/2021/11/12/n_16842907.shtml

35 The phrase "Comrade Stalin Strait" was coined by the father of the Russian hydrogen bomb, Andrey Dmitrievich Sakharov, a Soviet nuclear physicist, dissident, Nobel laureate, and activist for disarmament, peace, and human rights. Sakharov posited a number of hypothetical plans for launching a preemptive strike against the United States. One proposed detonating 50- to 100-megaton thermonuclear charges just off the two coasts of the U.S. The result would be giant tsunamis that would essentially wash away the United States to form the Stalin Strait between Canada and Mexico. See (2018, 17 March) "The Comrade Stalin Strait — Truth and Fiction" [original in Russian: "Пролив им. Товарища Сталина» правда и вымысел"]. https://fishki.net/anti/2539769-proliv-im-tovariwa-stalina-pravda-i-vymysel.html and (28 January 2022) "The Stalin Strait and other miracles of Andrey Sakharov" [original in Russian: "Пролив имени Сталина и другие чудеса Андрея Сахарова"]. https://sptoday.ru/2021_12_24/proliv-imeni-stalina-i-drugie-chudesa-andreya-saxarova/

36 Ilyushina, Mary. (2021, December 22). Putin threatens 'retaliatory military-technical' measures as standoff with U.S. and NATO over Ukraine escalates. *CBS News*. According to the

The need for a national ideology that centers on Russian Orthodox-Soviet values

17) Faced with a deadly challenge to the very sovereignty of its existence, Russia needed an official ideology of containment and revenge. "We want Russia to have an official ideology based on the teachings of the Russian Orthodox Church—an ideology that is adhered to no matter what.[37] We want this ideology to be the only—or at least the main—basis for the foreign policy of the Russian State"—the will of patriotic Constantinople.[38] Western analysts even came up with a name for this ideologeme—"strategic conservatism."

18) But the Kremlin is only flirting with Russian nationalism; it is afraid (like the liberal West) of its full implementation. The Russian people, on the other hand, need Russia to pull itself away from its "neutral" drifting pattern and make a shift from its "universal" demagoguery back to the Russian religious root meanings of its imperial existence. This way it would be able to achieve a reconciliation of its national Russian identity—the foundation of the Russian state. The Russia that was a great power is being rebuilt with a great sense of urgency, its greatness based on its historical traditions: the ideas of having a special path, an identity, sovereignty, and a specialness peculiar to Russian civilization (as the poet said: "It has a specialness: You can only believe in Russia").[39] The domestic and international landscape is being adjusted in accordance with Orthodox-Soviet values.

19) Thanks to the Kremlin's advancing of its own brand of "strategic conservatism," according to the experts of the Washington Center for Strategic and International Studies,[40] it seeks to achieve the following goals:

---

piece, Russian president Putin warned that if the U.S. and NATO do not halt what Moscow considers aggressive actions along the country's border with Ukraine, Russia would respond with "retaliatory military-technical" measures. Further, Putin emphasized that "we have every right to do so." https://www.cbsnews.com/news/russia-ukraine-war-news-putin-retaliatory-military-technical-measures/

37 In Dr. Vertlieb's original Russian text, he uses an Old Church Slavonic turn of phrase meaning, very loosely, "come hell or high water" or "no matter what happens," [in Russian, иже не прейдеши].

38 Here Dr. Vertlieb's original Russian text refers to the city of Constantinople by its ancient Russian name— Tsargrad [in Russian, Царьград] or "Tsar's City.

39 Taken from an 1866 poem by poet Fyodor Tyutchev: "You can't understand Russia with the mind//You can't measure it with a common yardstick://It has a specialness—//In Russia you can only believe." [In Russian: Умом Россию не понять//Аршином общим не измерить://У ней особенная стать—//В Россию можно только верить».] https://en.wikiquote.org/wiki/Fyodor_Tyutchev

40 Conley H., Ruy D., Stefanov, R., and Vladimirov, M. (2019, May 1). *The Kremlin Playbook*

- Reduce pro-Western sentiment in target countries;
- Strengthen support for Russian political actions (at home and abroad) and legitimize the Kremlin's narratives;
- Undermine support for EU membership among Member States and reduce support for EU and NATO membership in candidate countries;
- Keep the countries of the post-Soviet space in Russia's sphere of influence;
- Undermine internal cohesion, sovereignty, and possibly territorial integrity in a way that supports the interests of the Kremlin (e.g., Bosnia);
- Remove or weaken the leadership of the Ecumenical Patriarchate (which is seen as preventing the unification of the Orthodox world under Russian leadership); and,
- Lift sanctions (collateral and long-term benefits) and push Western governments to take Russia's political interests into account.

20) According to Western analysts, this is what the "arrows on the offensive map" of the new Russian ideology—"strategic conservatism"—look like. Time will tell: Is Russia really "concentrating" on its overdue national revival under the ancient guiding banners of Christ as a "symbol of victory over death and the devil," and not doing so in vain for the sake of an immediate gain (to prevent Ukraine from joining NATO and to successfully predict the result of the current competition for succession to the throne—the 2024 presidential election)?

21) For Russia, the primary task is to restore its original geopolitical "code" in terms of its worldview, which is a set of key ideas that Russians believe about their place in history and the world, their foreign policy strategy, and their national priorities. The Russians are trying to take the Chinese approach to politics: Without fail, in all agreements, the Chinese require "duiden"—parity in relations, measures, and steps. In accordance with the spiritual concept of "yin and yang" ("chaos and order"), they require a ranking of the entities involved—and an end to any chaotic activity. This approach is evident in Russia's requirement for NATO to return to its pre-1997 positions—before the beginning of the alliance's self-propulsion towards expansion—[and Russia's demands] to prevent the West from its military coopting of Ukraine—to

2. Center for Strategic & International Studies. 116 Pages. ISBN-10: 144281111. ISBN-13: 978-1442281110. For additional information, see also Conley H., Mina, J., Stefanov, R., and Vladimirov, M. (October 27, 2016) *The Kremlin Playbook: Understanding Russian Influence in Central and Eastern Europe.* (CSIS Reports) Paperback, 86 Pages. ISB-10: 1442279583, ISBN-13: 978-1442279582.

observe the "red lines" of the Russian Federation. And for this, Russia needs a parity of forces with the West, achievable by possessing, as a retaliatory counterthreat, an asymmetric mega-weapon capable of destroying the United States and Europe—as well as an accompanying conservative-value ideology of Russian victory.

22) Genuine conservative values are like a locomotive's flywheel, its source of power. They are necessary for any social system in order for it to prevent an ultra-liberal "maidanism"[41] from destroying the very machinery of the state. There is a fierce outbreak of ethnic nationalism taking place. Chauvinism has inflated this hate to the point that there is loathing for another people simply for what they are (Trumpists, Russians). It has the hallmarks of an impending coup d'état in the USA, even a second civil war. There is a change underway heading in the direction of a one-party system (with all its medieval "Soviet" consequences: "who is not with us is against us") — a trend towards left-fascist-leaning socialism. The political camps are still the same: the left (sinistram)—the sinful, bad—and the right (iustum)—the true, correct. The right, being more conservative, has relied on traditional American ethics: individualism, self-sufficiency, hard work, independence from the state, being law-abiding, and possessing equal rights before the law. The left, with its Marxist views, has been drawn to social values: collectivism, herding, subordination to superiors, the primacy of the state over the individual, and economic equality. These two approaches have manifested themselves in the two parties: the Republican Party and the Democratic Party. So far, Antifa and the BLM, under the slogan "Down with the Police," are the manifestations of a barricade-free pluralism of societal orientations.

23) The language of political correctness is "newspeak" à la George Orwell: robbery has become "redistribution of wealth"; a black criminal has become "a victim of racism"; talent has become "white privilege"; "white" has come to mean "racist." "We," as J. Fraden[42] states, "are waiting for the Second Civil War. It will be the only way to save the country. It will not only be necessary, but also legal. Let me remind you that for this reason, 244 years ago, the

41 "The Maidan Uprising was a wave of demonstrations and civil unrest in Ukraine, which began on 21 November 2013 with large protests in "Maidan Nezalezhnosti" [In Ukrainian, Independence Square] .... The protests were sparked by the Ukrainian government's sudden decision not to sign the European Union-Ukraine Association Agreement, instead choosing closer ties to Russia and the Eurasian Economic Union." https://en.wikipedia.org/wiki/Euromaidan

42 Jacob Fraden is an electronic engineer, inventor, entrepreneur, educator, artist, and writer. He has authored short stories in Russian and is a contributor to the on-line blog "American Thinker"—a daily online magazine dealing with American politics from a politically conservative viewpoint.

U.S. Founding Fathers wrote in the Declaration of Independence: "But when a long train of abuses and usurpations, pursuing invariably the same Object evinces a design to reduce them under absolute Despotism, it is their right, it is their duty, to throw off such Government, and to provide new Guards for their future security." The Russian national consciousness, which was damaged twice, in 1917 and 1991, also yearns for a conservative restoration of the "olden times."

24) Everything comes full circle. As the Eurasianist Pyotr Savitsky argued, "no matter who would have won the Civil War—the 'whites' or the 'reds'—all the same, Russia would be opposing the West, all the same it would be a great power, all the same would be creating a Great Empire."[43] If that is so, then the neo-Byzantine model of statehood is fitting for the creation of Eurasian Russia based on a combination of the religious values of Orthodoxy and the values of the Empire, headed by an Autocrat (as an option, someone with those functions, be it a Leader or General Secretary).

25) The concept of Alexander Prokhanov's "Fifth Empire" is based[44] on the Byzantine "symphony of authorities,"[45] where the Church and the monarchy/"leader of the people" closely cooperate in a single social liturgical work—universal salvation. May Monomakh's "state of truth"[46] shine—the triumph of justice, salvation, goodness. True conservatism can only be the faithful

43 "Petr Savitsky was a Russian pioneer of the so-called 'structural geography,' and was the first to propose...a geopolitical vision of 'Eurasia,' an entity which...is neither Europe nor Asia, but the 'place of development' of the Russian Empire...." https://www.researchgate.net/publication/298446006_Russia-Eurasia_according_to_Savitsky

44 Yasmann, Victor. (2006, November 3). Russia: The Fiction and Fact Of Empire. *Radio Free Europe/Radio Liberty broadcast (13:20 GMT)*: "The book *Symphony of The Fifth Empire*, is a collection of essays calling on Russia's elite, liberals, and patriots alike to unite to construct a new Eurasian empire—a successor to the Soviet Union and Tsarist Russia." In discussing his work, Prokhanov states, "One can see signs of an emerging empire almost everywhere...in events such as the building of new types of ships and submarines...launching the new 'Bulova' missile...or the construction of the North European Gas Pipeline."

45 Poliantseva, A. V. (2016). "Symphony" of the Authorities in Byzantium, and in Russia as the Successor of the Byzantine Empire. [In Russian: "Симфония" властей в Византии и России как преемницы византийской империи] *Polythematic Online Scientific Journal of Kuban State Agrarian University*. According to the article, "The Byzantine 'symphony of authorities' refers to the fact that in Byzantium, the authority of the Emperor played a major role in strengthening the Orthodox Church. The Church...developed and highlighted the official doctrine of the divine origin of the Imperial power. [Therefore] in the Byzantine Empire a perfect model of Church-state relations—a 'Symphony of the authorities' was formed." A dual-language version can be found at http://ej.kubagro.ru/2016/04/pdf/99.pdf

46 Vladimir Monomakh was a 12th century Grand Prince of Kiev revered for his institution of legal reforms which helped to establish a "state of truth"—rule of law—that extended legal protections to the lower-class.

preservation of the Divine Truth, and not the opportunistic tasks of the ruling "elite." The state and the Church are united only if they fulfill a common root national mission—that together with the people, they carry out their own transformation and the transformation of the world. Metropolitan Hilarion predicts a great spiritual future for the Russians, applying to it the truth found in the gospel that "the last shall be first."

26) The "stern face" of Russian imperial conservatism gets unfairly disparaged, with detractors likening it to "Stalinist Satanism" and classifying Putin and Dugin's "Eurasian" projects[47] as "right-wing extremist intellectualism in a neo-authoritarian Russia." "Hyperconservatism" is generally a boogeyman for liberals. "Ultimately," political scientist A. Malashenko asserts, "this 'hyper' leads to the collapse of the state, quite possibly pressured by the anger from the street and under the battering of extremism."[48] Historian A. Minakov also believes that conservatism in Russia always correlates with a strong, centralized, powerful hierarchical leadership structure. For in Russian conditions, only such a power can provide the mobilization of both material and human resources needed to conduct numerous wars. Survival is possible only with an extremely powerful centralized power, perceived by the people as a blessing and identified with what is now called the civilizational code. The essential values behind Russia's fundamental, spiritual system for preserving its way of life are an inextricably linked variation

47 Dixon Klump, Sarah. (2017, March 17). Russian Eurasianism: An Ideology of Empire. *Woodrow Wilson Center Press. Kennan Institute*. "Eurasianism rejects the view that Russia is on the periphery of Europe, and on the contrary interprets the country's geographic location as grounds for a kind of messianic 'third way.'" Further, according to Eurasianist Alexander Dugin, "Geography, not economics, is the pivotal cause of world power, and Russia, by its intrinsic physical location, [provides] a prime global role." The article goes on to say, "Dugin argued that Eurasian empire will be constructed on the basis of denying the common enemy, the USA and its liberal values.... The rigour of Dugin's influence in Russia has seen a steeping increase under Putin's rule...[however] many scholars have argued that Dugin's sway over Putin is highly overstated. In fact, there is no plausible evidence to show a direct link between Putin and Dugin, but the palpable ideological similarities shown by both of them regarding specific issues denote that Dugin has left some influence over Vladimir Putin's mind." For information on Eurasianism and Dugin, respectively, see https://www.wilsoncenter.org/publication/russian-eurasianism-ideology-empire and Amarasinghe, Punsara. (2020, April 8) Alexander Dugin's Neo-Eurasianism in Putin's Russia. *Modern Diplomacy* online at https://moderndiplomacy.eu/2020/04/08/alexanders-dugins-neo-eurasianism-in-putins-russia/

48 Interestingly, in 2017, in an online Belorusian discussion forum, political scientist and historian Aleksey Malashenko asserted that "without liberal influence, conservatism will turn into obscurantism that is disastrous for society, while liberalism, without regard for its opponent, will 'enrage,' and its radicalism will acquire extremist features.... Liberalism and conservativism must 'sing a duet.'" https://politring.com/region/906-aleksey-malashenko-konservatizm-i-liberalizm-mogut-pet-tolko-duetom.html

of the triad Spirituality, Sovereignty, and Sobornost[49]—the content make-up of the "Russian idea."

27) The church schism of the 17th century violated the integrity of this Triad. A sharp conflict arose between the secular and spiritual authorities which ended with the assertion of the primacy of the tsar's power over the power of the patriarch. Archpriest Avvakum did not change the ways of the Old Believers (such as the use of two fingers in rites and in the sign of the cross). His zealous adherence to tradition was inherited by the guardians/poshvenniks/conservatives[50]—supporters of unreformed Orthodoxy.

28) Putin's understanding of conservatism is not that of Admiral A. S. Shishkov,[51] who was one of the first to talk about the fact that the westernized Russian upper class had turned into a sort of special group of people living within the confines of a large population which has preserved genuine Russian values. While maintaining the gap in the Russian Federation between the "gold yacht Russians" and the "down-and-out" ones, a mortal battle of the "two Russias" cannot be ruled out—a battle between the original "projections" [for Russia] by the liberals/Westerners and by the conservatives/custodians of "imperial aspirations." I expressed this premonition of a collision in my article "Overcoming Systemic Threats to Russia's National Security":[52]

49 In Russian religious philosophy, "sobornost" (in Russian, собо́рность) refers to a spiritual community marked by a free spiritual unity in both the group's church and secular life. Khomyakov emphasized the unity of the Orthodox church which he juxtaposed with protestant individualism, or the "cult of individualism," found in the West. https://en.wikipedia.org/wiki/Sobornost and https://iphlib.ru/library/collection/newphilenc/document/HASH0178cc6530a5c64973161178/

50 The "guardians" here refer to those who adhere to the following belief: "One of the main concepts that made up the philosophy of Orthodox conservatism is the religious and political doctrine 'Moscow is the Third Rome,' which was the core of Orthodox conservatism." The "pochvenniki" [почвенники]—derived from the Russian word "pochva" [Russian: "почва" or "soil"]—refers to those who, like the Slavophiles of the 1860s, advocate for a merging of the educated class with the people ("soil"). *See Church and State in the Views of Russian Conservatives of the 19th Century. Religion in conservative public thought.* https://www-allistoria-ru.translate.goog/allis-488-6.html?_x_tr_sch=http&_x_tr_sl=ru&_x_tr_tl=en&_x_tr_hl=en&_x_tr_pto=sc

51 A. S. Shishkov (17154-1841) was a Russian writer and statesman whose intense nationalistic and religious sentiments made him a precursor of the Slavophile movement. He retired in disagreement with the early liberal reforms of Alexander I. Devoting himself to the promotion of Russian patriotism, he became a self-styled philologist, insisting that the Russian language be purged of foreign, especially French, influence. He was the author of "Discourse on Love for One's Country." https://www.britannica.com/biography/Aleksandr-Semyonovich-Shishkov

52 Vertlieb, Evgeny. (2007, February 3). Overcoming systemic threats to Russia's national security ["Преодоление системных угроз национальной безопасности России."]. Original Russian article published by the online newspaper "Forum.msk.ru." https://forum-msk.org/

"Speaking of the national security of modern post-Belovezhskaya Russia,[53] it needs to be clarified which of the two currently existing 'Russias' is under discussion. After all, the liberal pandemonium of the 1990s smashed the country into two unequal opposing 'Russias'—the oligarchic-mafia Russia and the Russia of the destitute. The fatherland, spiritually taken over and plundered by unjust privatization, has been actually split into the fiefdom of the super-rich and the pitiful existence of the destitute majority. Society has been torn into two antagonistic camps. Accordingly, each of these structures has its own set of ideological and worldview criteria for the truth, different needs, incomparable challenges and threats, prospects for the future, and scenarios for the strategic development of the country."

29) And just recently, V. Putin proposed his draft for a new state ideology—"healthy conservatism." Sociologist Karl Mannheim[54] believes that a special ideology was needed—one that serves to protect a defined political or social order from the external and internal challenges that threaten it. A discussion of the steps for creating a new state ideology follows.

30) The Belovezhskaya conspiracy, the Search for the "Russian Idea" in the Post-Soviet Era, and the steps for creating a new state ideology

31) The Belovezhskaya conspiracy destroyed the Soviet Union [and initiated] the newest phase of Russian history—first in the community of the CIS,[55] and then in the form of a broken-down, de-sovereign-ed, comatose Russian Federation—one that shone with the Shchedrin delight of the Foolovites[56] who freed themselves from themselves. (And

material/society/20484.html

53 Belovezha refers to the Belovezha Accords that ended the Soviet Union and established the Commonwealth of Independent States. See (2016, December 7) "History in the Making: The Agreement That Ended the Soviet Union." *The Moscow Times*. https://www.themoscowtimes.com/2016/12/07/history-in-the-making-the-agreement-that-ended-the-soviet-union-a56456

54 In the view of German Sociologist Karl Mannheim (1893-1947), social conflict is caused by the diversity in thoughts and beliefs (ideologies) among major segments of society that derive from differences in social location. https://www.britannica.com/biography/Karl-Mannheim

55 The Commonwealth of Independent States (CIS) [in Russian, Содружество Независимых Государств (СНГ)] is a regional intergovernmental organization in Eastern Europe and Asia. When the USSR began to fall in 1991, the founding republics signed the Belavezha Accords, declaring that the Soviet Union would cease to exist and proclaimed the CIS in its place. https://en.wikipedia.org/wiki/Commonwealth_of_Independent_States

56 This is a reference to the 19th Century satirical novel by Russian author M. Ye. Saltykov-Shchredin, *History of One Town* [История одного города]. The work is a farcical history of "Stupid Town" (Russian: Глупов) that follows the lives of "bungling" Russian "StupidTown-ites" (Russian: Глуповцы) for hundreds of years as they endure the violence and lunacy of their tyrannical rulers.

now, June 12th—as of 1994—has become the official holiday for celebrating the "independence of Russia." But independence from what? Was it a colony? Independence from the USSR?) The fatherland, torn apart in the "dashing 90s" by "criminal revolutionaries" ("the great criminal revolution" — the term used by film director S. S. Govorukhin[57]), was deprived of the ideology of the "Soviet Union"[58] (the Red Empire), by the implementation of the 13th Article of the Constitution of the Russian Federation which prohibits the establishment of any ideology—"any state ideology or any ideology at all that is mandatory." The comprador oligarchy ruling the Russian Federation was quite satisfied with the established collaborative-corporate system of the "pipe economy" with the policy of the "gas at a discount" agreement.[59]

32) But by God's providence and Putin's efforts, the coup d'état wound up "falling short of the goalpost" (my wording-E.V.). Russia is being reborn from the ashes and ruins. In a situation where there is an escalation in threats and challenges facing the Russian Federation, the issue of creating a national ideology has become relevant, the ideology being a harmonizer of society, an optimizer of the spirit of the people, and a veritable "TNT" of information-organizational weapons for use in the geopolitical confrontation of global national interests.

33) True, the Russian class of bosses certainly has something it can use to eliminate the threat of a "Birch Revolution"[60] in the Russian

57 "Russian film director Stanislav Govorukhin produced a documentary film about the two years following the dissolution of the USSR in 1991...The movie—also a book by Govorukhin—exposes the ex-Communist officials who became Russia's *nouveaux riches* by getting a leg up on amassing wealth when Gaidar decontrolled prices, as well as the mafia kingpins who became their fellow travelers to billionairehood through extortion rackets." See Govorukhin, S. (1993) *The Great Criminal Revolution* [In Russian: Великая криминальная революция]. Andreyevsky Flag Publishers. 126 Pages. ISBN 5856080262. Govorukhin produced a documentary film by the same name which is summarized by Douglas, Rachel. (1994, July 15) "Documentary film on Russian crime is presented in Washington." *Executive Intelligence Review*, 21(28), 142-145.

58 Here, in Dr. Vertlieb's original Russian text, he uses a derogatory, slang term for the former Soviet Union— "sovka" [совка].

59 Here, "gas at a discount" is used ironically: In the absence of a real state ideology respected by the people, Russians have to get satisfaction from an ideological surrogate like "discount gas." Similarly, to gain support of, say, Belarus, the Russian Federation—an anti-people state—has to sell gas cheaply to them. (E.V.)

60 "Birch Revolution" is a term used to refer to a threat to the Kremlin's power from pro-Western democratic opposition allegedly intending to overthrow Putin (or his protégé in the 2008 presidential election) with financial assistance from Western non-governmental organizations modeled on the "colored revolutions." The "predicted" revolution never took place. The term "birch revolution" first appeared in 2005 and apparently came from the Russian word for birch [in Russian, *berez* - берёз), a tree that symbolizes Russia. https://dic-academic-ru.

Federation. This, as characterized by philosopher Olga Malinova, is a set of "comparatively stable and recognizable systems of meanings."[61] But this methodology has been compromised: "There was an impact on those lacking any rights by the dictates of other members of society, and the potential became obvious for an 'abuse of power' to weaken the competitive chances of opponents up to limiting ideological pluralism by banning the expression of certain ideas in public spaces .... The ruling elite does not have the right to use state instruments of coercion to impose their own views as binding or to exclude the right to express other points of view."

34) From the standpoint of dealing with the irreconcilable opposition, it is true that the State Duma often adopts laws that are akin to laws of war "occupation." Many forms of public protest have been declared illegal. Take, for example, those calling for fines or forced labor for "insulting a representative of the authorities" (*Criminal Code of the Russian Federation, Article 319*). Such legal purists are terribly far removed from the people. Russia, thrown off the path of its ordinary way of life and even off its civilizational path by the Belovezha putsch,[62] still cannot fully identify itself and normalize itself under the Law. The project that was created—capitalism with a human face—is a failure in terms of the harmony of the masses: the oligarchy is too greedy for its super-profits, and the authorities too unwilling to earnestly grant "social guarantees" to its citizens. And the USSR-2 project still terrifies the imagination with nightmares of the GULAG.

35) Concerning the question of the interconnectedness of pluralism, state ideology, and personal freedom ... The liberal conservative jurist Boris Chicherin[63] considered political freedoms—the right to participate

translate.goog/dic.nsf/ruwiki/157992?_x_tr_sl=ru&_x_tr_tl=en&_x_tr_hl=en&_x_tr_pto=sc

61 Olga Malinova, Doctor of Philosophy, is a chief research fellow at the Institute of Scientific Information on Social Sciences (INIO) at the Russian Academy of Sciences. Malinova's "systems of meanings" refers to a mechanism used by the *nomenklatura* to control who occupies positions of power. For further information, see Dr. Vertlieb's "Project Putin-2024 in the Geostrategy of Confrontation and Internal Challenges." *Global Security & Intelligence Studies.* 6(2), Winter 2021, 210.

62 "Belovezhskaya putsch" refers to the signing of the Belovezha Accords that ended the Soviet Union and established the Commonwealth of Independent States. See (2016, December 7) "History in the Making: The Agreement That Ended the Soviet Union." *The Moscow Times.* https://www.themoscowtimes.com/2016/12/07/history-in-the-making-the-agreement-that-ended-the-soviet-union-a56456

63 Boris Nikolaevich Chicherin was a 19th century Russian jurist and political philosopher who proposed the theory that what Russia needed was a strong, authoritative government to persevere—but one with liberal reforms. https://en.wikipedia.org/wiki/Boris Chicherin

in state power—to be the highest development of personal freedom and its only guarantee: "As long as the government is independent of citizens, their rights are not guaranteed to be free from its arbitrariness: in relation to the state, a person is powerless."[64] The conservative-liberal concept is more organic—more suited—for taking into account the entire polyphony of the spectrum of different views in society. To harmonize society, the approach used in the cartoonish-Yeltsin demagogy of Leopold the Cat is clearly not enough: "Guys, let's live together!"[65] There is no declared "consensus" of the victims of theft and the thieves—the beneficiaries of the "priKhvatization."[66] The blatant social inequality between the "down-and-out class" and the "gold yacht class" is in no way consistent with the thesis of living in genuine harmony "in accordance with fair justice and the law."

36) For three decades, the post-Soviet ideosphere had been developing an extreme anti-statism, minimizing of the role of the state and propagandizing a program of godless enrichment instead one of conscience. Such a monstrous shift in the Russian paradigm was codified in the super-presidential Constitution.[67] Then the undermined strengths of an already minimized and weakened state provoked a series of terrorist attacks from 1991 (Chechnya and Dagestan) to 2004 (Beslan).[68] But Vladimir Putin then managed to reverse the "harmful megatrend" (regression, deregulation)

64 Chicherin, Boris Nikolaevich. (1866). Dissertation entitled "On the Representation of the People," Book 1 "The Being and Properties," Chapter 1 "Representation and Authority" [Russian: "О народном представительстве," Книга 1 "Существо и свойства народного представительства," Глава 1 "Преставительство и полномочие"]. https://www.rcoit.ru/lib/history/narodnoe_predstavitelstvo/18307/

65 "Leopold the Cat is a Russian/Soviet animated series about a cat named Leopold. He is very kind and wants to do only good, but the world is not only good; it is also evil. Two mice are constantly trying to hurt him, but the Leopold always smiles and says, 'Guys, let's all live together in harmony! [In Russian, Ребята, давайте жить дружно!]' The series was produced from 1975 to 1987." https://www.imdb.com/title/tt1074757/

66 Dr. Vertlieb coins a word here—"preKhvatization [приХватизация]"—that combines the words for "privatization [приватизация]" and "to seize [приХватить]" with the implication being that the privatization carried out following the dissolution of the USSR was essentially legalized stealing.

67 Amendments to the Russian constitution in 2020 included changes allowing Putin to run again for two more six-year presidential terms as well as conservative amendments such as constitutionally ensuring patriotic education in schools and placing the constitution above international law.

68 The dissolution of the USSR in 1991 was quickly followed by Chechnya declaring its independence. The Beslan school attack refers to the violent takeover of a school in Beslan, North Ossetia, Russia, in September 2004 by militants linked to the separatist insurgency in the nearby republic of Chechnya.

by strengthening his "vertical of power" and accelerating national self-identification.

37) In October 1994, S. Shakhrai and V. Nikonov published the "Conservative Manifesto"[69]—"conservatism with a Russian face." It contained the main postulates of classical conservatism, supported by quotations from W. Churchill to K. Leontiev. At the same time, an attempt was being made to create an ideological doctrine of "democratic patriotism" (V. Shumeiko, V. Kostikov).[70] The constituent elements of the "new ideology" centered on the concept of the formation of a political nation of "ethnic Russians" and included: general patriotic rhetoric that included the notion of Russia as a great state; statements about the need to reintegrate post-Soviet space, with the leading role played by Russia as the "first among equals"; and the resurrection of the inspirational phrase "united and indivisible Russia." However, the theory of "the new Russian nation" did not take root. But the real-world application of the inspirational phrase "united and indivisible Russia" in "restoring constitutional order in the Chechen Republic" wound up working positively. Russian statehood withstood the separatist challenge from Ichkeria.[71]

38) The period from 1996 to August 1998, according to historian Sergei Panteleev, was marked by an active search for a "national idea" designed to consolidate society. After a difficult victory in the presidential elections ("Vote or lose!"), which further exacerbated the ideological polarization of Russian society, B. Yeltsin, on July 12, 1996, initiated the process of developing a unified national doctrine by issuing instructions to find out "which national idea, which national ideology

69 Classical conservatism was founded by British statesman Edmund Burke and later developed by Russian thinkers of the 19th and 20th centuries such as Konstantin Leontiev. In post-Soviet Russia, the effort to revive this ideology was undertaken by Vyacheslav Nikonov and Sergey Shakhrai, who co-authored the Conservative Manifesto in 1994. The Manifesto was the ideological basis for activities carried out by the Party for Russian Unity and Accord (PRES) [Партии российского единства и согласия—ПРЕС]. See https://russia-direct.org/analysis/has-putin-pragmatist-turned-putin-ideologue and https://ru.wikipedia.org/wiki/Никонов,_Вячеслав_Алексеевич

70 Russian Federation Council V. Shumeiko and press secretary of the President V. Kostikov put forward the doctrine of "democratic patriotism." The doctrine was included in the President Yeltsin's 1994 address to the Federal Assembly "On Strengthening the Russian State." It assumed the formation of a political nation of "[ethnic] Russians," and proposed that Russia to become the "first among equals" in the Commonwealth of Independent States. https://cyclowiki-org.translate.goog/wiki/Государственная_пропаганда_в_России?_x_tr_sch=http&_x_tr_sl=ru&_x_tr_tl=en&_x_tr_hl=en&_x_tr_pto=sc

71 The Chechen Republic of Ichkeria is the unrecognized secessionist government of the Chechen Republic. The republic was proclaimed in 1991, which led to two wars with the Russian Federation. https://military-history.fandom.com/wiki/Chechen_Republic_of_Ichkeria

was the most important for Russia."[72] However, the crisis of August 17, 1998—when, for the first time in world history, a state defaulted on domestic debt[73]—put an end to the socio-economic, political, and ideological course that had been pursued since 1992.

39) The stage of development from September 1998 to the end of 1999 was the sublimation of the *"conservative wave"* on the crest of which "Operation Successor" was carried out—the transition of power without any upheaval or revolution. The conservative doctrine of E. Primakov[74]—the strategy of strengthening the country through the construction of a "socially-oriented market with state participation"—met the expectations of the authorities and society. The ideological niche of "enlightened" conservatism (the veiled triad of Count S. S. Uvarov[75]) took root because of its position on the need for a "national idea" which was understood to mean "patriotism, sovereignty, statehood, and social solidarity."[76] But it is one thing to proclaim a conservative-sovereign state course, and another to implement it. As the then gray cardinal of the Kremlin, Vladislav Surkov,[77] said, "we are, of course, unconditional conservatives, although we don't know what that is yet." Conservatism as a fashionable brand: D. Trump defeats H. Clinton (USA); N. Farage—D. Cameron (England); the Five Star Movement[78]— the Brussels Bureaucracy (Italy); V. Orban—the forces personified by D. Soros (Hungary). One issue of the influential magazine *Foreign Affairs* stated that "Nobody in America these days seems to want to be a

72 See (1996, July 13) Nezavisimaya Gazeta [Независимая Газета], Page 1.

73 The Russian financial crisis (also called the "ruble crisis" or the "Russian flu") hit Russia on 17 August 1998. It resulted in the Russian government and Russian Central Bank devaluing the ruble and defaulting on its debt. https://en.wikipedia.org/wiki/1998_Russian_financial_crisis

74 Yevgeny Maksimovich Primakov served as Prime Minister of Russia from 1998-1999, and Minister of Foreign Affairs from 1996-1998. Yeltsin fired Primakov on 12 May 1999, ostensibly over the sluggish pace of the Russian economy.

75 "In 1832, a slogan was created by Count Sergey S. Uvarov, Minister of Education...that came to represent the official ideology of the imperial government of Nicholas I ... and remained the guiding principle behind government policy during later periods of imperial rule." See *Orthodoxy, Autocracy, and Nationality.* Britannica. https://www.britannica.com/topic/Orthodoxy-Autocracy-and-Nationality

76 See (1999, December 30) Independent Gazette [Независимая Газета], Page 4.

77 Vladislav Surkov, former Assistant to the President of the Russian Federation, is known in Russia as "the gray cardinal" for his behind-the-scenes political machinations. Surkov, who stage-managed Russia's involvement in Ukraine, was fired by Putin in February 2021. https://foreignpolicy.com/2020/02/21/putin-fires-vladislav-surkov-puppet-master-russia-ukraine-rebels/

78 The Five Star Movement is a political party in Italy lead by Giuseppe Conte, former Prime Minister of Italy, from 2018 until 2021.

liberal—or even be known as one." And the conservative publication *National Review* printed an excerpt from a book by the philosopher J. Hazony under the heading "Liberalism as Imperialism."[79] And then in 2018, another book by this author on the same topic, *The Virtue of Nationalism,* was widely declared to be the most important publication of conservative thought since Huntington's famous book *The Clash of Civilizations.*[80]

Putin's Brand of Conservatism and Russian Philosophical Underpinnings

40) Such is the trend of the times. And Russia is no exception: It is moving from Westernism to patriotism, from radicalism to conservatism, from "free market" to statism, from ideological nihilism to a single national idea. V. Putin calls his ideology "healthy" or "reasonable" or "moderate" conservatism. Well-known Russian ideologist and Eurasianist Alexander Dugin wrote in a collection of essays entitled *Liberalism is a Threat to Humanity*:[81] "Putin, intuitively seeking to preserve and restore Russia's sovereignty, came into conflict with the liberal West and its globalization plans but did not formalize this into an alternative ideology either." Was he really going to "shape" some kind of extensive anti-Western ideology? Hardly. His conflict with the liberal West is not of an antagonistic nature. "Unlike Boris Yeltsin, V. Putin is not a radical. Unlike the Soviet leaders, he is not a dogmatist."[82]

41) Putin's conservatism is the ideology of preserving society as it is at the moment (of course, that is, with the prospect of immanent development). It is clear: he seeks evolutionary changes but "without creating shockwaves." Because society is polarized, it is impossible to please both opposing sides. But the

---

79 Hazony, Yoram. (2018, September 6). Liberalism as Imperialism. *National Review.* March 18, 2022. https://www.nationalreview.com/2018/09/liberalism-as-imperialism-dogmatic-utopianism-elites-america-europe/ and Hazony's *The Virtue of Nationalism* (2018, September 4) Basic Books. 304 Pages. ISBN-10: 1541645375 and ISGN-13: 978-1541645370.

80 Huntington, S. (1996). *Clash of Civilizations and the Remaking of World Order.* Simon & Schuster (2011, August 2). 368 Pages. ISBN-10: 1451628975 and ISBN-13: 978-1451628975.

81 Dugin, Aleksandr. (2014, April 23). Liberalism is a Threat to Humanity. *Blagodatnyy Ogon' Provoslavnyy Zhurnal.* [Original published in Russian: Либерализм–угроза человечеству. Благодатный Огонь Православный журнал.] https://blagogon.ru/biblio/636/print

82 Medvedev, Roy Aleksandrovich. (2010). *Vladimir Putin. To Be Continued.* Vremya Publishers. 304 Pages. ISBN: 978-5-9691-0264-4. [Original in Russian: Владимир Путин. Продолжение следует.] According to Medvedev, "Putin is not only a European, but a Westerner as well... pro-market as well as a socialist, a liberal and a conservative who is formulating a new Russian ideology.... He wants to include the sensible ideas from all different modern ideologies and take into account for the new Russia, all the values from its former epochs...including the Soviet era."

president's choice of whom to please is his. His Valdai speech about ideology is based on the "philosophy of inequality" of the philosopher N. Berdyaev.[83] But the speech does not say anything about an intention to organize public life to provide justice for each and every member of society. This means that what is actually "conserved" is this: a deliberately unfair structure for Russian society. And it is unrealistic then for such a government to expect love from the people.

42) But what has to be taken into account is the reflex reaction of the people— the visceral conservatism of the Russian consciousness. This is a phenomenon of being drawn to tradition, like an "ethnogravitational constant" (my term-E.V.). For example, Dostoevsky's character Falalei constantly sees a dream about a white bull.[84] Another example is from the politician V. S. Chernomyrdin: "Whatever public organization we create, we wind up with the CPSU [Communist Party of the Soviet Union]."[85] And isn't Russian "conservatism" itself functionally the same? Indeed, in Russian political science, it is synonymous with the concept of the "monarchy" and understood to mean autocracy. Therefore, the "guardianship" of the true values of the Russian root system must, a priori, be pro-imperial, pro-Orthodox, with the primacy of justice, conscience, honor, valor—and Truth above all. Such is the need to support the organic existence of the Russian people.

43) Putin's ideological statement [in October 2021 at the Valdai Discussion Club] is not extreme, but

83 In October 2021, Putin delivered a speech to the 18th Plenary Session of the Valdai Discussion Club in Sochi, Russia, in which he made reference to Russian philosopher Berdyaev, thereby "appearing to...encourage personal creativity, dialogue and engagement rather than a self-serving ideology and paternalism...." But Berdyaev's philosophy was one of personal freedom. His 1917 work, *The Philosophy of Inequality*, "voiced a powerful critique of societal myths and mentalities that lead to a crushing totalitarian control over life." See https://www.barnesandnoble.com/w/the-philosophy-of-inequality-nicholas-berdyaev/1122025853 and for further details on the Valdai address, https://asiatimes.com/2021/11/the-optimistic-conservatism-of-putins-valdai-address/ and https://www.themoscowtimes.com/2021/10/22/putin-rails-against-monstrous-west-in-valdai-speech-a75373

84 Zholkovsky, Alexander. (1994). *Text Countertext: Rereadings in Russian Literary History*. Stanford University Press. Page 333. "Dostoevsky spoofs the traditional recurrence of the 'same old dream' not only by iteration but also by the dream's very content: The Russian idiom skazka pro belogo byka/bychka ("a tale about the white bull" or, in Russian, "сказка про белого быка/бычка") means 'endless repetition of one and the same thing'—always being drawn to tradition, for example."

85 Viktor Stepanovich Chernomyrdin was Prime Minister of Russia from 1992 to 1998. Interestingly, he was known for his malapropisms and syntactic errors. Many of his sayings became aphorisms and idioms in the Russian language, one example being the expression "We wanted the best, but it turned out like always" [in Russian: "Хотели как лучше, а получилось как всегда."] https://en.wikipedia.org/wiki/Viktor_Chernomyrdin

broadly humanitarian, without significant innovations: "The conservative approach is not one of mindless guarding, not a fear of change, and not a game of playing to hold much less retreating into one's own shell. It is primarily a reliance on a time-tested tradition—(yes, but which one exactly? Imperial? recent Soviet? or that of today's thieves? -E.V.)—reliance on the fundamental rejection of extremism as a way of action."[86]

44) As stated earlier, Putin's ideological concept is based on Berdyaev's "philosophy of inequality." If this is any indication of an attempt to inject a system of inequality into Russian society, then it is difficult to shake the thought of the people's lawful, in God's eyes, opposition to such an abuse of power. And a rebellion in Russia, according to Pushkin, is "resolute and merciless."[87] The urgent need for total justice that Russian society demands can no longer be ignored by the populist regime which allows only cosmetic changes to its ugly, anti-people nature, rather than what is expected—a dismantling of an inherently alien and rotten structure. Although the Kremlin assures that there will be no return to 1937,[88] TV commentator V. Solovyev time and again has proposed to the leadership that he be appointed the sword of retribution of the security apparatuses—Chief of the Death-To-Spies [organizations].[89] Everything is not so unambiguous in this "kingdom of distorted mirrors."

86 Russian President Vladimir Putin's address to the plenary session of the Valdai Discussion Club was reported, in part, online by the Russian business daily newspaper, *Vedomosti* (2021, October 21). In his address, Putin announced the principles of healthy conservatism which is at the heart of Russia's approaches. [In Russian online: Путин заявил о принципах здорового консерватизма в основе подходов России.] https://www.vedomosti.ru/politics/news/2021/10/21/892404-printsipah-zdorovogo-konservatizma

87 This reference is taken from Chapter 13 of Russian writer A. S. Pushkin's novella *The Captain's Daughter*, Chapter 13 [in Russian, Капитанская дочь.] One translation of the original reads, "God forbid we should see a Russian rebellion, senseless and merciless!" [In Russian, " Не приведи Бог видеть русский бунт, бессмысленный и беспощадный!" https://dic.academic.ru/dic.nsf/dic_wingwords/2402/Русский

88 "On July 31, 1937, one of the most terrible documents in history was signed—the secret operational order of the NKVD No. 00447, which marked the beginning of the events known as 'Yezhovism.' Yezhov was head of the People's Commissariat for Internal Affairs (NKVD) -- the secret police -- under Stalin. See Krechetnikov, Artem. (2017, July 31) "Stalin's Strike: 80 years ago the Great terror started in the USSR." *BBC Russian Service, Moscow.* [Original in Russian: Сталинксий удар: 80 лет назад в СССР начался Большой террор.] https://www.bbc.com/russian/features-40756213

89 Here Dr. Vertlieb uses his own portmanteau for "Chief of the Death-to-Spies [organizations]"—Glav (Chief)+ SMERSh (acronym for Death to Spies)+suffix for a person—in Russian: "Глав+СМЕРШ+евец" or ГлавСМЕРШевец. SMERSH actually refers to the motto of an umbrella organization established for three counterintelligence Red Army agencies formed in about 1942—SMER(T') or DEATH+ShPIONAM "TO SPIES."

45) And if N. Berdyaev appeals to you, then take great care: this existentialist philosopher is a "great sower of confusion" and can lead you astray. More suitable for the development of the Kremlin's new ideology seems to me to be the founder of Russian political science, Professor of Moscow University B. N. Chicherin (even though N. A. Berdyaev declared him an "enemy of democracy"). Chicherin defends the constitutionalist, political-legal ideal. His guiding principle—the *Course in the Science of Government*[90]—was his desire to "reconcile the principles of freedom with the principles of power and law." This was the basic postulate of the Chicherin program of protective liberalism, the main political slogan of which is "liberal measures and strong power." In addition, in a technical way, he more skillfully combines the concepts of "inequality" with "justice": "Equality in the possession of material goods follows from the requirements of justice just as little as the equality of bodily strength, mind, and beauty follow from one another." Equality of rights (formal) cannot be replaced by equality of possessions (of material goods). Few people in Russian political culture have established the ontological nature of inequality so convincingly unless it is the philosopher and diplomat Konstantin Leontiev for whose conservatism Leo Tolstoy[91] and Fyodor Dostoevsky were only "pink Christians"[92]—individuals who were not adequately working to counter the tendency of the Russian people to

90 "In his *Course in the Science of Government* (1894) [in Russian: Курс государственной науки], Chicherin formulated four asks for the state: to provide security, protect civil rights, uphold the moral order through the rule of law and justice, and to pursue the public good." See (2009) Chicherin and Shipov: two competing visions of local self-government and central representation from the 1890s to the early 1900s. *Australian Slavonic and East European Studies*, 23 (1-2), 39-56.

91 "The Tolstoyan movement is a social movement based on the philosophical and religious views of Russian novelist Leo Tolstoy. Tolstoy denied that there was any actual 'movement' but was gladdened by the fact that groups of people were declaring complete agreement with his views.... 'Tolstoyans' identified themselves as Christians, but who did not generally belong to an institutional church. Tolstoy was a harsh critic of the Russian Orthodox Church, leading to his excommunication in 1901." https://en.wikipedia.org/wiki/Tolstoyan_movement

92 Orekhanov, Georgy and Pushchayev, Yury. (2012, September 26). Pink Christianity (Part 1): The Lonely Thinker vs. Tolstoy and Dostoyevsky. *FOMA Magazine*. [Original in Russian: Розовое христианство (часть 1): Одинокий мыслитель против Толстого и Достоевского.] Konstantin Leontiev called Tolstoy and Dostoevsky "pink Christians" because of the questioning way the writers dealt with Christianity in their writings. Leontiev predicted the falling away of the Russian people from Orthodoxy and believed these authors were not helping to prevent this. The way the authors dealt with Christianity in their works was therefore only "rose-colored" or "cosmetic." Some believe that the term stemmed from the fact that in that time, "rose water" was a very popular facial lotion. https://foma.ru/rozovoe-xristianstvo-ch1.html

fall away from Orthodoxy. Deep churching, i.e., completely leading an Orthodox life, prompted Leontiev to become a monk.

46) "President Putin," I wrote in 2007, "is probably familiar with the liberal-conservative concept of Boris Chicherin, who formulated a political principle that is very suitable for the current government in Russia—liberal measures and strong power." Characterizing Berdyaev, Chicherin said, "He was a rare statesman in Russia, very different in this way from both the Slavophiles and the left Westerners .... He accepts the empire, but wants it to be cultural and to absorb liberal legal elements." These are points consonant with Putin's.

47) And one more pencil stroke in the blueprint that Chicherin draws for devising an ideology: Berdyaev was indeed confused. "Conservatism," writes philosopher Chicherin, "is not what prevents moving up and forward, but what prevents going back and down, towards chaos."[93] By context, chaos is inherent in the state of affairs that exists before one arrives at conservatism. Putin's "chaos" is a hyperbole, used in a figurative sense, an assimilation of the subversive (terrorist) tendencies of the masses toward chaos. Berdyaev's impulse is of a different mix: Yes, "the pressure of chaotic darkness from below" exists (this is true), but not due to anarchist-protestants, but, rather, as a result of "the inherent primordial-sinful [Original Sin-based], animal-chaotic element found in human societies." Like some sort of pre-societal phantoms. And the main thing for Berdyaev is this: "Chaotic formless darkness in of itself is not yet evil, but only a boundless source of life." Putin's idea of everyday "extremism" does not follow from Berdyaev's ontological "evil." Generally, things are not going well with the Russian Federation's top expert on "evil" although the Institute of Philosophy of the Russian Academy of Sciences did spend 742,000 rubles on research on hell and evil.[94]

48) Evil is not something that is there from the beginning, but, rather, something that arises after the fact. Confucius testifies to this: "Evil has no independent cause in the universe." Evil could not be created by the benevolent Heavens as an independent element of the world. It stems from the violation of order (which does not mean "public

93 In his address to the Valdai Discussion Club in October 2021, Putin cited precisely these words of Chicherin's about conservatism. See McDonough, Tom. (2021, November 5) The optimistic conservatism of Putin's Valdai address. *Asia Times*. Online at https://asiatimes.com/2021/11/the-optimistic-conservatism-of-putins-valdai-address/

94 "In recent months, the Institute of Philosophy of the Russian Academy of Sciences has ordered several studies in the titles of which the words 'hell' and 'evil' appeared. The latter took 742 thousand rubles.... All this is within the law,' the institute said." https://financialnerd.com/the-ras-institute-spent-742-thousand-rubles-to-study-hell-and-evil/

order!"-E.V.), that is, from the violation of good by a failure to understand the heavenly order. We inject disorder into the world, destroying its original harmony; we create chaos in it, thereby violating and destroying its original order. This is how misfortunes and troubles appear. This is how evil appears. Thus, evil is the result of an upset in the world's balance of order. Evil is an imbalance in the universe. Berdyaev's idea is that chaos correlates organically with order. Order is a positive, primary category of the universe (as in Chinese cosmogony), and not something that falls in the range of "mass disorders" (disharmony, rebellion).

49) So it's precisely the right time to encourage the acceptance of libertarian conservatism as part of a new [Russian] ideology—a right-wing political movement that seeks to combine libertarian and conservative ideas, that is, one that tries to develop the idea of preserving traditions and maintaining a conservative course of development while maintaining the individual freedom of every single person. "Sobornost" allowed a person to see an himself as an individual or, in the words of A. I. Klibanov,[95] Christianity taught the internal sovereignty of the individual. Based upon its general shape and essence, Putin's ideologeme is "liberal conservatism" while the very need of Russia, as a besieged fortress, is something different: it is "the strategic conservatism of Victory."

50) Conservatism can be a component of various worldview combinatorics. So an "ideology of order and protection"[96] is questionable. After all, "order," in the slogan "anarchy is the mother of order," is not legitimate when referring to "protection." The same is true for power. And if it is a "whitewashed autocracy"? Conservatism for Russia is a historically natural and mentally sound consolidator of the people and a prerequisite for establishing a new balance of social intentions —a balance between the "icon" (order) and the "axe" (rebellion).

51) Not believing in a voluntary change to the "unchanging" course of the country, the supporters of the core

95 Aleksandr Ilyich Klibanov (1910-1994) was a Russian historian, religious scholar, and author of numerous scholarly works. For a detailed biography, see https://www.pravenc.ru/text/1841323.html

96 Nemensky, Oleg. (2013, June 5). Protection [in Russian: Охранительство]. *Political News Agency*. Nizhny Novgorod "Ideology of order and protection" refers to an ideology calling for the preservation of order and the maintenance of the status quo. In an online article addressing the use of the Russian word "protection," Nemensky claims that the word "cannot be translated into other languages...[but] most often it is understood as an ideological orientation towards maintaining the *existing state of affairs*, especially with regard to state power." https://apn--nn-com.translate.goog/analytic/okhranitelstvo/?_x_tr_sl=ru&_x_tr_tl=en&_x_tr_hl=en&_x_tr_pto=op,sc

Russian political orientation—the people's monarchist adherents—rely on a "national military dictatorship" to save the Fatherland. The "putsch scenario" of a change in power is also being written about in the West, for example, in Catherine Belton's book, *Putin's People: How the KGB Took Russia, and then Took on the West.*[97] One thing is clear: without a change in the paradigm of the current course of the Russian Federation—the "nationalization" of the very logic behind Russian thinking and the strengthening of the pro-Russian dominant element—the country will lose not only sovereignty, but also the vital resources for the reproduction of the nation. The degradation and depopulation of the population of the Russian Federation is a demographic catastrophe with geostrategic consequences: With a numerically small population of 145, 975, 300 people, it is difficult to preserve the Motherland from Kaliningrad to Vladivostok.

52) The "main foundation" of the guiding ideology should be the assimilation of the axiom: the Russian idea—imperial in essence and "the creator of national self-consciousness, culture, and religious providence—as the destiny of the nation." Its categories—Spirituality, Sovereignty, and Sobornost—are triune, modeled on the Holy Trinity and the triad of Good, Truth, and Beauty. And remember both that a liberal is not a boogeyman, and that a conservative is not [a source of] salvation if he does not live in Russia. As Herzen[98] spoke of the Westernizers and the Slavophiles, they "looked in different directions" but "the heart beat the same." The exuberance of youth such as theirs grows wiser with age, they live without clashing with one other: "He who was not a liberal in youth has no heart, he who in maturity has not become a conservative has no brain."

53) In terms of any concrete reconciliation of the ideological camps within the Russian world, one biographical fact comes to mind: In the late 1970s, philosopher Pyotr Boldyrev and I were the first in the world abroad to try to reconcile "liberals" and "conservatives" with one other. In the joint article, "Solzhenitsyn and Yanov"—published in Dovlatov's *New American*[99]—instead of

97 Belton, Kathryn. (2020, April 2). *Putin's People: How the KGB Took Back Russia and Then Took On the West.* Farrar, Straus and Giroux, 640 pages. ISBN-10: 0374238715, ISBN-13: 978-0374238711.

98 Aleksandr I. Herzen (1812-1870) was a Russian writer and thinker known as the "father of Russian socialism." Among other things, he fought for the emancipation of the Russian serfs, and after that took place in 1861, he escalated his demands regarding constitutional rights, common ownership of land, and government by the people. https://en.wikipedia.org/wiki/Alexander_Herzen

99 The article appeared in (1981, March 24) *The New American*, New York, 59, 36. Sergey Dovlatov, unable to publish in the USSR, resorted to underground *samizdat* as a means for circu-

the usual "either-or," we combined the opposite poles, replacing this wasteful, fruitless disjunction with a reassuring "and-and." We asked ourselves the question: Is there any sort of "unity of opposites" discernible in the historical concepts of these opponents or in their interpretations of Russian history? And with all the apparent and undeniable dualism, are there any obvious irreducible elements the two share? It turned out that both "progressives" (liberals) and "intuitionists" holding unchanging values in history (i.e., conservatives—supporters of "substance of the people," "community," and "the soul of the people") ideologically "lapsed" into the Russian "liberal conservatism" or "protective liberalism" of B. Chicherin, who was highly valued by both the liberal, P. Struve,[100] and the conservative, I. Ilyin.[101] The credo of conservative (concrete) liberalism can be expressed something like this: not a "liberalism in general terms" but one that includes national-cultural traditions; not a provincial, constrained, national "conservatism" (protection) but one based upon universal cultural values. Sometimes one can get lost among these various ideological azimuths. For example, in England, liberalism is eight centuries old and began with the "Magna Carta Libertatum." Therefore, the liberal values of individual freedom are more traditional. Does this mean that adherence to the precise values of liberalism turns out to be the most extreme form of conservatism?

Seeking a Middle-of-the-Road Solution

54) For the rhetorical question we posed then: Isn't it time for the modern Russian opposition (with its historiography) to pay closer attention to this third, "middle" path—the drifting of those intent on confrontation from a position of confrontation to synthesis, where we would get the most of out of both [liberalism and conservatism]? A healthy balance of conservatism and liberalism serves to correct both of these spiritual tendencies, neutralizing their

lating his works. He was expelled from the country and became a prominent figure in the New York émigré community. He served as a co-editor of *The New American*, a liberal Russian-language émigré newspaper. He finally received recognition as a write when he was printed in *The New Yorker* magazine. https://en.wikipedia.org/wiki/Sergei_Dovlatov

100 Petr B. Struve (1870-1944) was a Russian political economist, philosopher, historian, and editor. He started out as a Marxist, later became a liberal, and after the Bolshevik Revolution, joined the White movement. From 1920, he lived in exile in Paris, where he was a prominent critic of Russian communism. https://en.wikipedia.org/wiki/Peter_Struve

101 Synder, Timothy. (2018, March 16). Ivan Ilyin, Putin's Philosopher of Russian Fascism. *The New York Review*. Ivan Ilyin (1883-1954) provided a metaphysical and moral justification for political totalitarianism, which he expressed in practical outlines for a fascist state. Today, his ideas have been revived and celebrated by Vladimir Putin. https://en.wikipedia.org/wiki/Peter_Struve

extreme forms of expression—the descent of conservatism into "obscurantism," and of liberalism into "extreme progressivism." Establishing defining lines among social groups is more complicated than with analogs from natural science. Ideologies, like metals, are subject to transmutation and are good neighbors with their own kind, the metalloids. The classification is elementary with metals: unpaired d-electrons give life to transition metals. But the secrets of social groups are revealed by devilish master keys: neural networks now have been taught to determine the political views of a person. Artificial intelligence can identify supporters of liberalism and adherents of conservative views from photographs. It turns out that liberals more often look directly into the camera lens, expressing surprise. Liberals cannot stand even the smell of conservatives. It was experimentally deduced that conservatives are more squeamish than liberals. The conclusion is far-reaching: perhaps the fact that liberals have little to no distaste for anything, it pushes them to change. It pushes them to hold protests. Or even seek revolutionary changes. On the other hand, conservatives turn out to be happier than liberals: The more liberal a person is, the more unhappy he is. And vice versa. The phenomenon is explained by the difference in their understanding of justice. What prevents liberals from being happy is the feeling that the difference between the rich and the poor is too great, that public goods are distributed somehow wrongly. Stemming from this, they, as a rule, are unhappy even in their personal lives.[102]

55) The new ideological conglomeration of the conservative and liberal heritage —of all that is the best of each—inevitably includes the basic values of "classical" liberalism such as: a) the absolute value of the human individual and the natural ("from birth") equality of all people; b) the existence of certain inalienable human rights such as the right to life, personal freedom (while not infringing on others' freedom), and justice; c) the creation of a state

102 Numerous articles address studies made of liberal and conservative "brains"—physical differences related to varied political thinking. For information on artificial intelligence and predicting political orientation, see Yirka, Bob (2021, January 14) AI algorithm over 70% accurate at guessing a person's political orientation, *Tech Xplore,* https://techxplore.com/news/2021-01-ai-algorithm-accurate-person-political.html, and for information on political neuroscience, Denworth, Lydia (2020, October 26). Conservative and Liberal Brains Might Have Some Real Differences, *Scientific American*: "Social scientists who observe behaviors in the political sphere can gain substantial insight into the hazards of errant partisanship. Political neuroscience, however, attempts to deepen these observations by supplying evidence that a belief or bias manifests as a measure of brain volume or activity—demonstrating that an attitude, conviction or misconception is, in fact, genuine.... Partisanship does not just affect our vote; it influences our memory, reasoning and even our perception of truth." https://www.businessinsider.com/psychological-differences-between-conservatives-and-liberals-2018-2

based on general consensus in order to preserve and protect the natural rights of man; d) the rule of law as an instrument of social control and "freedom within the law" as the right and opportunity, as stated by John Locke, "to live in accordance with a permanent law common to everyone in the society and to not be dependent on the fickle, indefinite, unknown autocratic will of another person"; and, e) the ability of each individual to spiritually progress and seek moral perfection. "Participating in the economic life of society," Adam Smith, the 18th century philosopher, maintains that "every person, in addition to satisfying his own interests, involuntarily contributes to the fulfillment of common interests, because they are nothing more than 'the sum of the interests of individual members of society.'"

56) The main features of conservatism are the following: "The preservation of the ancient moral traditions of mankind; respect for the wisdom of ancestors; rejection of radical changes in traditional values and institutions; the conviction that society cannot be built according to speculatively designed schemes; happiness is impossible without harmonious relations with society"[103]—notions of ancient philosophers and of conservative E. Burke. In modern conservatism, different groupings are united by common concepts, ideas, and ideals. Although conservatism is traditionally identified with the defense of the social status quo, a characteristic feature of the modern conservative renaissance has been the fact that it was the neoconservatives and the "new right" who initiated the changes aimed at restructuring the existing order. Therefore, the churning of turmoil both on the left and on the right is not ruled out. There is more in common between the extremes than between the centrists.

57) Apparently, this duality is natural. After all, liberalism initially has a conservative element. In his theory of the social contract, the master of liberalism, John Locke, means that the people do not fully transfer their power to the state; they only delegate it for the protection of the whole society. In other words, by creating the state, people sought

103 Author unknown. (2013, January 25). Liberalism and conservatism—from confrontation to synthesis. [In Russian: Либерализм - от противоборства к синтезу.] *REFSRU*. "The concept of 'conservatism' was first used by the French writer Chauteaubriand. The history of conservatism...begins with the Great French Revolution, which challenged the very foundations of the 'old order,' all traditional forces, and all forms of domination by the feudal aristocracy. It was from that time that two classical forms of conservatism originated: the first, which comes from the French thinkers J. de Maistre and L. De Bonand; the second - from the English thinker Edmund Burke." Burke's brand of conservatism "was destined for a long life...Burke's book *Reflections on the Revolution in France* marked the emergence of conservatism as a social current...." https://www-refsru-com.translate.goog/referat-16144-2.html?_x_tr_sl=ru&_x_tr_tl=en&_x_tr_hl=en&_x_tr_pto=sc

to safeguard their civil interests, to protect their freedom. As N. Andreev states, "It is about freedom of conscience, thought, the protection of one's legal status, the fulfillment of tax obligations sanctioned by the state and codified in legislation, etc. In light of these things, the liberals defend the need for each government to adopt a constitutional act that enshrines the rights and freedoms of the individual and citizen." The Slavophiles, who profess inner truth as the primary regulator of social elations, were appalled by the regulated legalism of the "agreement." According to their conviction, the Autocracy and the People are united by their common professed Orthodoxy. As I. Aksakov[104] wrote: "That is the whole essence of the union of the Tsar with the people, that they have one divine moral basis of life, one God, one Judge, one law of the Lord, one truth, one conscience." The activities of both the Earth (the people) and the Tsar are based on Divine Will and Truth. Consequently, according to the Slavophile doctrine, there can be no talk of any contractual principles for the emergence of power. The best form of government for Russia is the Monarchy, which has provided true freedom.

58) The ruling nomenklatura is terribly far removed from the people. If it was the extreme "liberals" who ruined Russia in 1917, why are such people still in power?! In a situation where people and government coexist as if in parallel realities, and the "servants of the people" serve the ruling comprador oligarchy, often adopting clearly anti-people legislation, most citizens of the Russian Federation are waiting for a real pro-people policy, implemented no matter what the party emblems, if only by "the leader's own down-to-earth" interests and incorruptible principles.

59) Anything that threatens Russian identity should not be sanctioned by the authorities unless in a limited one-time use—how Emperor Paul allied with heretic Catholics in opposition to the French Revolution. And then, with perestroika, the principle of catholicity ("sobornost") was, in fact, eliminated, and a synod was not declared as the main authority of the church, but, rather, an administration of the patriarch and bishops. Metropolitan Hilarion, after meeting with Pope Francis, said: "There is a very great prejudice against Catholics among the Orthodox people, and we must in no way risk the unity of our churches and the peace in our churches, therefore we must move forward in our relations with Catholics with all speed possible." Is it a coincidence

104 Ivan Sergeyevich Aksakov was a noted 19th century Slavophile, controversial journalist, newspaper publisher, and proponent of Pan-Slavism. His brother, Konstantin Sergeyevich, was one of the founders and principal theorists of the Slavophile movement. https://www.britannica.com/biography/Ivan-Sergeyevich-Aksakov

that when Russia is gathering itself, Patriarch Kirill warns the "bosses" against tyranny? Or Yeltsin's irremovable system of "checks and balances" is being triggered? It is also "tradition." Conservatism is being faithful to the original Plan of God, that is, to the Orthodox religion and to the corresponding ideology of how to organize the earthly life of the state. This is the "blossoming complexity" of attempting to lay out the specifics of a civilization.

60) Attempts to conserve the ruling regime in the Russian Federation with state "conservatism"—completely in opposition to the Russian national tradition and to the recognition of the Russian people as a conciliar entity—are fraught with upheavals and a degradation of the country. For an imperial monarchic mindset, state conservatism is unacceptable, being protective both from the excessive claims of the West and from its own people—legalized by the oligarchic "multinational" constitution and the increase in punitive "extremist" articles in the Criminal Code. Unfortunately, the current church leadership often understands its conservatism as the priestly service to any authority which supposedly does not exist if it is "not from God." But under the current conditions, a liberal-Maidan-type rebellion could bring rulers to power in Russia who are even more hostile to Russian tradition. The "Ideology of Victory as a National Project"[105] is designed to counter the ideology of "inclusive capitalism" of the new globalist order named after Klaus Schwab.[106]

---

105 Averyanov, Vitaly. (2021, October 28). The Ideology of Victory as a National Project. [In Russian: Идеология победы как национальный проект.] *The S. P. Kurdyumov ANO Center for Interdisciplinary Research.* According to Averyanov, "A civilization awakening from a traumatic hibernation is in dire need of an integrating ideology. In 2000-2020 the authorities made certain efforts to restore traditional patriotism around the image of the victorious country.... One can even say that modern Russia already has an ideology — the ideology of Victory.... In our Victory, all imperial tendencies really merged...feuds fell silent...pre-war contradictions were blunted. It was a grandiose, mystical, religious in spirit Victory.... [Now there is an] urgent need to focus on what and over whom we have to win a new Victory, the successor Victory.... [Also needed is] the revival of historical optimism, the mythology of Great Development under the slogan "Give back the dream to the people." https://spkurdyumov.ru/future/ideologiya-pobedy-kak-nacionalnyj-proekt/

106 Schwab, Klaus. (2019, December 2). What Kind of Capitalism Do We Want? *Project Syndicate.* World Economic Forum founder and Executive Chairman, Klaus Schwab is the originator of "'stakeholder capitalism,' a model he first proposed a half-century ago.... [It] positions private corporations as trustees of society, and is clearly the best response to today's social and environmental challenges.... [To] ensure that stakeholder capitalism remains the new dominant model...the World Economic Forum is releasing a new "Davos Manifesto," which states that companies should pay their fair share of taxes, show zero tolerance for corruption, uphold human rights throughout their global supply chains, and advocate for a competitive level playing field...." 'Stakeholder capitalism' appears to be synonymous with 'inclusive capitalism' which "is fundamentally about creating long-term value for all stakeholders [providing] equality of opportunity...equitable outcomes...fairness across generations...and fairness in society." https://

God's Providence itself obliges Russia to play the role of that same Restrainer, about whom the apostle Paul spoke (2 Thessalonians 2:7),[107] in this battle against world evil.

61) Today, the Russian government positions itself as an adherent of conservative values but acts more often as a pragmatist and a "distilled liberal." Writer Alexander Prokhanov,[108] when asked what is wrong with liberalism, answers, "Russia's experience of interacting with liberals ends completely tragically for the country—Russia disintegrates. Liberals can be colorful and attractive but their presence in power leads to the fact that they destroy the state without offering any other foundations. Russia is collapsing and in order to restore it, we must bring back the imperial form of government at a huge cost—at the cost of losing historical time and losing people. Left and right domestic conservatives are "great-state-power advocates" for whom a strong state is one of the key values. Russia is currently still struggling with the legacy of the "third turmoil" of 1991.[109] These are lost territories, and economic problems, and "tacit support for the Washington Consensus."[110]

62) Juliette Faure, a French researcher of contemporary Russian conservatism, writes about this in her dissertation about dynamic conservatism, which is being created in an ideology that is based on the synthesis of tradition and modernity.[111] The author quotes the main ideologist of "dynamic conservatism," Vitaly

www.project-syndicate.org/commentary/stakeholder-capitalism-new-metrics-by-klaus-schwab-2019-11 and https://www.inclusivecapitalism.com/what-is-inclusive-capitalism/

107 The verse reads: "For the mystery of lawlessness is already at work; only He who now restrains will do so until He is taken out of the way." New King James Version.

108 "Alexander Prokhanov is a prize-winning Russian novelist and, as editor of the weekly newspaper Zavtra [Завтра], a leading figure in Russian imperial patriotism" according to a description of Edmund Griffiths' book *Aleksandr Prokhanov and Post-Soviet Esotericism* found at http://cup.columbia.edu/book/aleksandr-prokhanov-and-post-soviet-esotericism/9783838209630

109 Legacy of the Third Turmoil of 1991 apparently refers to the aftermath of the failed 1991 Soviet coup d'état attempt by communist hard-liners to take control of the country from Mikhail Gorbachev, who was Soviet President and General Secretary of the Party. https://en.wikipedia.org/wiki/1991_Soviet_coup_d%27état_attempt

110 The Washington Consensus is a set of ten economic policy prescriptions considered to constitute the "standard" reform package promoted for crisis-wracked developing countries by Washington, D.C.-based institutions such as the International Monetary Fund, World Bank, and U.S. Department of the Treasury. https://en.wikipedia.org/wiki/Washington_Consensus

111 Faure, Juliette. (2019, January 28). The idea of tradition at the heart of the politics of the contemporary Russian regime: a "dynamic conservatism"? [In French: L' idée de tradition au cœur de la politique du regime russe contemporain: un "conservatisme dynamique?"] *Centre de recherches internationales (Sciences Po, CNRS)*. This work constituted her Masters thesis. Faure is a PhD candidate in political science at Sciences Po Paris and Visiting Fellow at Harvard University's Davis Center for Russian and Eurasian Studies. https://spire.sciencespo.fr/no-

Averyanov,[112] who claims that the task is "to create a centaur from Orthodoxy and innovation, from high spirituality and high technology. This centaur will represent the face of Russia in the 21st century." Apparently in no way can it be done without the "centaur"—a frightening symbiosis of wisdom and courage in the person of the ancient Slavic deity Kitovras.[113] For, according to Zbigniew Brzezinski, one cannot win the Cold War without having an alternative ideology: "To be a military adversary of the United States on a global scale, Russia will have to fulfill some mission, implement a global strategy and ... acquire an ideological basis." While this seems "unlikely" to Brzezinski, Russia is trying different options to "save itself and the world." He is imagining a future system—a meritocracy (leadership by the best) under the auspices of a specially created official Strategic Council of the Russian Federation, equipped with powerful "ideological weapons." That, however, should not hamper, as the philosopher I. Ilyin says, "the soul of the Russian people which always seeks its roots in God and in his earthly phenomena —in truth, righteousness and beauty."

63) It is clear, Greeks of present day, neo-Byzantinism "is all ours." But from the standpoint of a temporary "reverse perspective"[114] (D. S. Likhachev's term),[115] those "front eyelids" are the most important thing for us: there is the root cause of today's "lack of unity" within the Russian Federation, both in the mental and territorial sense. Without a strict, rigorous assessment of what happened to the Motherland—from the Nikon schism to

tice/2441/51gf1nlacp8hhaepbqffd19i09

112 Faure, Juliette. (2021, April 3). 'Dynamic conservatism': A Russian version of reactionary modernism. *Ideology, Theory, Practice.* Faure notes that "...after the fall of the USSR, Prokhanov's modernism became commonplace among the members of the younger generation of conservative thinkers...[who] framed their views in the post-Soviet context, and regarded technological modernity as instrumental for the recovery of Russia's status as a great power." https://www.ideology-theory-practice.org/blog/dynamic-conservatism-a-russian-version-of-reactionary-modernism

113 In this retelling of Efrosin's 15th century philosophical tale, the author writes, "The biblical king Solomon, who was building the Holy of Holies (the famous Jerusalem temple), needed the advice of Kitovras." Ultimately, "thanks to the advice of Kitovras, the construction of the temple is successfully completed." Versions of the entire legend are available at https://en.inbel.org/3408-the-legend-of-how-kitovras-was-taken-by-solomon.html and https://sourcebook.stanford.edu/text/efrosins-tale-solomon-and-kitrovas

114 Using reverse perspective in iconography means transferring the point of presence from the viewer to the icon itself...the icon is looking at us. Chechko, Daria. (2020, March 5). The Reverse Perspective in Iconography. Article posted online by *The Catalogue of Good Deeds.* https://blog.obitel-minsk.com

115 Dmitry Likhachev (1906-1999) was a Russian medievalist, linguist, and GULAG survivor. President Boris Yeltsin often sought his advice on cultural and historical issues.

the execution of the parliament—it is impossible to develop a national ideology for the Russian people.

Putin's War against Ukraine: Between a Rock and a Hard Place—Strategic Conservatism?

64) With the collapse of the Soviet Union, the "Soviet" ideology, as a remnant, was done away with. However, given the conditions of the now exacerbated civilizational clash between historical antagonists, the Russian establishment needed a patriotic ideology to consolidate the nation. Thus, the ideology of "strategic conservatism" was born.[116] It was put to the test by the "second winter war"[117]—being fought for the "denazification" of Ukraine (a term referring to the eradication of Nazi ideology—Entnazifizierung). The question arises: If the purpose of this special operation *is the destruction of the [modern-day version of the] "Nazi regime" of Bandera,*[118] then why is the Kremlin conducting separate negotiations with the "Ukies"[119] without

116 Putin's strategic conservatism actively seeks "to influence religious and traditional views to further its own agenda...and reflects the idea that political and cultural preferences can be used as tools of influence." Putin uses the Russian Orthodox Church, Orthodox oligarchs, and intellectuals who support either him or the ROC's efforts to achieve his political aims... In addition, Russia positions itself as a defender of the traditional order and conservative values—the political and cultural embodiment of the Third Rome. The Kremlin amplifies this message through U.S. and European conservative networks." See Conley, Heather A. & Ruy, Donatienne. (2022). The Kremlin Playbook 3. *Center for Strategic and International Studies.* ISBN: 978-1-5381-7045-8 (pb); 978-1-5381-7046-5 (eBook). An article summarizing the work can be found online at https://www.csis.org/features/kremlin-playbook-3

117 Dr. Vertlieb is coining his own term for the current Russian military actions against Ukraine. The "Winter War" is the term the Finns use to refer to the Russo-Finnish War in 1939-40. The underlying cause for the war was Soviet concern about Nazi Germany's expansionism.... Finland itself was not a threat to the Soviet Union, but its territory, located strategically near Leningrad, could be used as a base by the Germans. The Soviets initiated negotiations with Finland that ran intermittently from the spring of 1938 to the summer of 1939, but nothing was achieved. Finnish assurances that the country would never allow German violations of its neutrality were not accepted by the Soviets, who asked for more concrete guarantees. In particular, the Soviets sought a base on the northern shore of the Gulf of Finland... [but the] Finnish government...felt that accepting these terms would only lead to further, increasingly unreasonable, demands. http://countrystudies.us/finland/19.htm

118 Putin claims that the current operation against Ukraine and its leadership is a repeat of the Soviet military actions against Stepan Bandera (1909-1959), i.e., an elimination of "pro-fascist" forces. Bandera was the leader of the revolutionary faction of the Organization of Ukrainian Nationalists, which, along with its partisan army – the Ukrainian Insurgent Army–strove to eliminate all ethnically non-Ukrainian elements from Ukrainian soil (including Jews, Russians, Poles, Gypsies, etc.). For a certain period of time, Bandera collaborated with Germans in the hope of achieving this goal. See (2015, January 29) The Success of Russia's Propaganda: Ukraine's "Banderovtsy." *Cambridge Globalist.* http://cambridgeglobalist.org/?p=573

119 Although the term may refer to ancient tribes called the "Ukry" [Russian "Укры"] who were the forerunners of the modern-day Ukrainian people, another use of the term is to refer to

first completing its assigned task?! Is Putin retreating, having run up against unexpected (an intelligence miscalculation) strong enemy resistance? Or is it a tactic to achieve victory "by other means" (to blunt the vigilance of the enemy by exchanging the stick for the carrot)? After all, the initial application of the word denazification is widely contextual: it implies not only the cleansing of the *whole* of Ukraine, *but also wider* (Poland? Transdnestria?[120] The Baltic States?).

65) What made Putin enter into a Khasavyurt 2?[121] Is it only the interests of the oligarch Abramovich[122] and others like him looming behind the draft of a separate agreement? It is dangerous for the Kremlin to lose its patriotic facade in the eyes of the people who are expecting the "capture of Kiev" (so that the sacrifices will not have been in vain). The ruling elite is rushing between Scylla and Charybdis—between two bad options for the outcome of the operation. The Kremlin now finds itself in the position of Zugzwang—as in chess, being forced to make a move even though any move at all will lead to a deterioration of its position.

Evgeny Alexandrovich Vertlib / Dr. Eugene Alexander Vertlieb
President of the International Institute for Strategic Assessments and Conflict Management (MISOUK-France); executive editor of the Western Policy Forecasting Department of Slavic Europe (Munich); executive member of the Lisbon-Vladivostok Initiative (Paris)

Ukrainians in a derisive manner. The latter usage is what is intended here. It is used to reflect the attitude of the Russian ruling elite toward the people of Ukraine.

120 Transnistria, officially the Pridnestrovian Moldavian Republic (PMR), is an unrecognized breakaway state internationally recognized as part of Moldova.

121 Between 1994 and 2000, Russia waged two wars in Chechnya. The Khasavyurt Accord was an agreement signed in August 1996, that marked the end of the First Chechen War, but it failed to preclude a second war. "Khasavyurt 2," then, refers to the on-going peace talks between Russia and Ukraine.

122 Russian billionaire oligarch Roman Abramovich accepted a Ukrainian request to have him assist in peace talks between Russia and Ukraine. Abramovich was sanctioned by the EU and UK...over his alleged links to Russia's President Vladimir Putin, which he denies. But [Ukrainian President] Zelensky has reportedly asked the U.S. to hold off from sanctioning Mr. Abramovich, arguing he could play a role in negotiating a peace deal with Moscow. See Roman Abramovich suffered 'suspected poisoning' at talks, an 11 March 2022. BBC article online found at https://www.bbc.com/news/world-europe-60904676

Dr. Evgeny Alexandrovich Vertlib/Dr. Eugene Alexander Vertlieb is a Russian-born dissident holding U.S. citizenship and currently living in France. He received a B.A. at Leningrad (Sankt-Peterburg) State University, a PhD at the University of North Carolina, and completed a postdoctoral internship at the Russian Academy of Public Administration in the Russian Federation. Dr. Vertlieb has held a variety of positions including Professor at the Marshall European Center for Security Studies, Germany, and at the Institute of International Relations (U.S.). He is an author of several books and articles on geopolitics and comparative cultural studies. Dr. Vertlieb has also served as an independent political adviser to the Committee on International Affairs and International Relations, Kyrgyzstan. He was nominated for the Chingiz Aimatov Gold Medal for his contribution in formulating key philosophical elements reflecting Kyrgyz ethnicity for inclusion in the "Concept of National Security for Kyrgyzstan." Additionally, for his active and fruitful involvement in furthering military-diplomatic cooperation between the United States and Mongolia, he was awarded a Certificate of Honor of the Mongolian Ministry of Defense signed by the Mongolian General Staff Chief, General Tsevegsuren Togoo. He is currently President of the International Institute for Strategic Assessments and Conflict Management (IISACM-France); Executive Director of the Western Policy Forecasting Department for Slavic Europe, (Munich, Germany); Executive Member of the Lisbon-Vladivostok Initiative (France), and Head of the Department of International Relations of the world Christian organization "Blagovest Media International" (Brussels).

Dennis T. Faleris received a B.S. from the University of Michigan and a master's degree in Russian Linguistics from Georgetown University. For more than thirty-five years, Mr. Faleris worked as an instructor, translator, senior intelligence analyst, and intelligence production manager at the National Security Agency. His career centered around Soviet/Russian military issues as well as a variety of transnational issues. He currently resides in Annapolis, Maryland, with his wife, Kathleen.

Идеология «стратегического консерватизма» в имперской перспективе России

> «Несчастна страна, в которой всякий консерватизм сделался постылым и насилующим... То в стране готовится революция (философ Н.Бердяев).
>
> «Консерватизм без агрессивной имперской идеи превращается в проповедь мещанского конформизма, исходящего из логики – лишь бы не стало хуже» (историк А.Минаков).
>
> «Дмитрий Медведев не меньший – в хорошем смысле слова – русский националист, чем я» (В.Путин).

1) Христианские религиозные конфессии - православие, католицизм и протестантизм вышли из единой духовной основы: прарелигии. В 1054 произошел окончательный раскол церквей, восточной и западной: на восточно-православную и западно-католическую. Этот великий раскол стал главной причиной многих межцивилизационных войн.

2) В 1653 Русь присоединила себе Левобережную Украину, на территории которой действовал греческий обряд, распространённый также на Балканах и Ближнем Востоке. Русский обряд отличался от греческого. Например, иначе писалось имя Христа — Исус, а не Иисус, крестились двумя пальцами, а не тремя. Унификация русского и греческого обрядов объединила бы всех славян — восточных и балканских — и положила бы начало Великой Греко-Российской империи. Патриарх Никон и осуществлял церковную реформу от имени царя Алексея Михайловича. Однако внесение изменений в богослужебные книги и некоторые обряды в целях их унификации с современными греческими вызвало церковный раскол на «аввакумовцев» -старообрядцев и «никонян». Отголоски той духовной междоусобицы отчасти наблюдаются в конфронтации идеологий нынешних российских консерваторов и либералов. Современные староверы - носители традиционной русской народной религиозности vs. либералы-западники.

3) Русская культура духовно целостностна и нацелена на истину как живую онтологическую сущность мира. Никоновцы потревожили фундаментальные основы русского мировоззрения и миросозерцания. Смена обряда богослужения и структурные изменения в административной

церковной иерархии нарушали органичную связь Бога и Церкви. Ведь до тех новаций истинно верным считалось что только «двуперстие» (старообрядческое крестное знамение) отображает «истинную догматику христианского Символа Веры - распятие и воскресение Христово, а также два естества во Христе - человеческое и Божественное». В трёхперстном (новообрядческо-никоновском) же крестном знамении, в аспекте догматики содержания получалось искажение подлинного смысла: будто на Кресте была распята Троица! Церковной клятвой и постановлениями Стоглавого собора 1551-го закреплялась сохранность двуперстия: «Аще ли кто двемя персты не благословляет якоже и Христос, или не вображает крестнаго знамения, да будет проклят».

4) Этот «инновационный постмодерн» духовной трансформации менял само понятие греха. Дезориентированная паства, выбитая из традиционной колеи вероисповедания, теряла этическую уверенность в распознании «правого» и «грешника». На Страшном Суде «поставит Господь праведных одесную Себя (по правую руку), а грешников — ошуюю (по левую руку). Стояние грешников по левую руку означает и положение руки при совершении крестного знамения на левое плечо». Возникал у прихожан храма Господня подспудный страх не угодить бы из-за возникшей неразберихи к «еретикам безблагодатным». Патриарха Никона, заявлявшего о себе «Я по телу русский, а по душе грек», нарекли антихристом.

5) К единству расколотое православие возвращалось долго. По фольклорно-этническим повериям и традициям: сперва сращивали разъятые части, потом одухотворяли живой водой и уж затем – возвращали душу. На поместных соборах 1918 и 1971 Русская церковь признала равноспасительность старых обрядов. Единоверческая церковь вернула старообрядцев во Вселенскую Церковь. «Единоверие» как одна вера - Вселенской Православной Церкви. Митрополит Антоний (Храповицкий) восстановил истину о патриархе Никоне. С отстоянием тезиса «Византия – праматерь России» (чему помог убедительный фильм митрополита Тихона «Гибель империи. Византийский урок»), христианский космополитизм Никона был реабилитирован как «апостольский неовизантизм». Патриарх Никон, оказывается, и отстаивал тогда ныне востребованную в России идею «симфонии властей» («двуглавие» - духовной и светской) во главе с православным царём-самодержцем, помазанником Божьим.

6) Объединению православных ветвей духовенства способство-

вало противостояние обеих конфессий католичеству. Онтологично противоборство этих антиподов. Как утверждает философ-богослов В.В. Зеньковский, католичество, довольствуясь лишь внешней атрибутикой благочестия, не приняло учение Христа о внутреннем обращении человека к правде и любви. Тем самым отвергло Благовестие Христа о свободе (тогда как православие «во истину свободно во Христе»), а следовательно в глазах ортодоксального подлинноверия католичество «утратило истинно христианское начало».

7) Зато явилась идея теократии (Бог плюс управление), которая будучи приложенной к социуму - породила соблазн подчинить человечество единой власти (глобализм «мирового правительства»). Все беды отсюда. Как полагает В.Зеньковский, христианство, реформированное католицизмом, по духу «социалистично» - в смысле насильственного приведения людей к социальному «раю» материального благополучия. А от невольных последствий католической реформации христианства - «католического социализма» - рукой подать до революционного социализма с человеческим лицом: «тот хаос свободы, хаос аморализма, в котором пребывает современное человечество, создан тем, что католичество отвергло Христово учение о свободе». Модернизация внутри церковной жизни католицизма явилась катализатором революционно-позитивистских социальных потрясений. В «Легенде о Великом Инквизиторе» романа «Братья Карамазовы» Ф.М.Достоевского обыгрывается католическое утверждение, что человечество неспособно к христианской свободе. Потерявший веру в Христа инквизитор хочет разрешить социальную проблематику без Христа, но с помощью отошедшей от заветов «больной» католической церковной организации. Смысл этой аллюзии доказать примат необходимости «вселить в души идеал красоты» над призывами социалистов: «Накорми, тогда и спрашивай добродетели!».

8) Католичество, как полагает славянофил А.Хомяков, изменяет началу свободы во имя единства, а протестантство - наоборот. В этом концепте, только православие осталось верным духу первохристианства, являясь гармоничным сочетанием единства и свободы в принципе христианской любви. Католичество прониклось рационализмом, отвергнув соборное начало; протестантство лишь развивает католический рационализм, ведущий от единства к свободе. Запад, констатирует культуролог Н.А.Нарочницкая, - это свобода «от чего» (отсутствие ограничений), а Россия - свобода «для чего» (зачем нужна свобода).

Православие - это свобода служения христианской добродетели «во имя Отца и Сына Святаго Духа».

9) Если так, в православии сохраняется вся правда свободы, данной человеку, но преодолевается её хаос. Православию не нужно «уединение», как «эготичное развитие личностного начала». Русско-православный путь целительный: он предполагает, что социальные противоречия разрешаются не через насильственное навязывание человечеству счастья (потребительство, гедонизм, успешность, эгалитарный прогрессивизм), а посредством примирения всех и всего в лоне Церкви. Соборное православное сознание воцерковляет жизнь во Христе. Это и есть тот положительный идеал, который воодушевлял мыслителя Достоевского и который им понимался не как внешнее подчинение Церкви всей жизни (так представляет католичество), но как свободное и внутреннее усвоение жизнью христианских начал во всех её формах бытия. Русская всечеловечность способствует осуществлению Христовых заветов на земле. Через покаяние вернулся к Бого-этическому преображённому себе. Чем глубже пал человек - тем значительнее подвиг его нравственного воскрешения. Страдание - во искупление содеянного греха-зла. Если оно воспринято искренне и глубоко, свободно явлено и по-настоящему выстрадано - моральное исцеление возможно. Таковы духовные скрижали русского национального миропонимания.

10) Русская традиционно-религиозная модель мировосприятия базировалась на абсолютизированном противостоянии полюсов добра и зла. Русское сознание крайностно - «всё или ничего». Оно не любит середины (там серая чертовщина), бескомпромиссно по своей сути. Компромисс рассматривается как беспринципность, неумение проявить волю, настоять на своём. Нет прагматике «золотой середины»-целесообразности. Отрыв свободы от Креста - «грешный рай» - нравственный коллапс для русского сознания, незамутнённого ревизионизмом адаптации. Первозданная чистота православия обречена на «стязания с латиною». Ибо «верующий разум» - истинно Боговерный и вольный во Кресте. Православие сознаёт себя подлинным, нереформируемым христианством, тогда как реформация католицизма невольно способствовала революционизации сознания европейцев. Миссия православия –охранять дух в нравственной первозданной чистоте, а мировой порядок – в Божественной гармонии с правдой и справедливостью.

* * *

11) Для русского соборного сознания неприемлемы не только западные фундаментально-мировоззренческие новации, но и «антихристовы» реформы императора Петра Великого: за их беспощадность его нарекли «первым большевиком». Несмотря на всю технологическую пользу Отечеству, он нанёс народу русскому непоправимый вред: отменил патриаршество, подчинил Церковь государству и «приостановил» русскую святость. Прорубив окно в Европу, он надломил русскую консервативную традицию, замутил идентичность, изменил национальный код. Казус с «бородой» обкромсал консервативную «длинноволосатую» нормативность - обернулся разрушением традиционной российской иерархии ценностей. Пусть и усилил имперскую державность заимствованной у Европы атрибутикой. «Он хотел,- как подметил философ Жан-Жак Руссо, - сначала создать немцев, англичан, когда надо было начать с того, чтобы создавать русских. Он помешал своим подданным стать когда-нибудь тем, чем они могли бы стать». Император Пётр I решил преобразовать Россию в Голландию (и флаг нынешний в РФ – такой же чуждой масти триколор) ценой превращения государства «в игрушку нескончаемых перемен», ведущих к революции. С позиции консервативного видения, он подорвал «дух народный», то есть самые основы самодержавия, нравственное могущество государства. В связи с чем историк Н.М.Карамзин в Записке «О древней и новой России» констатирует: «Мы с приобретением добродетелей человеческих утратили гражданские. ...Мы стали гражданами мира, но перестали быть, в некоторых случаях, гражданами России. Виною Пётр».

12) Карамзин вменяет в вину Петру многое, а главное - создание европеизированного правящего слоя, который, по сути дела, перестал быть русским. Он обвиняет его в этом гигантском болезненном социокультурном расколе, разрыве между верхами и низами. «Вменение Петру создание социокультурной пропасти между верхами и низами, чреватой революцией. И это вменение, пожалуй, одна из главных скреп русской консервативной идеологии до 1917 года. Да и после» - констатирует историк Аркадий Минаков, добавляя: Пётр «верхний слой денационализировал. Сделал его космополитическим. А значит - малопригодным для решения тех проблем, которые стоят перед русскими и Россией».

* * *

13) Как Россия возвращалось к себе – своему корневому архаическому вероисповеданию? Древнерусский основательный «крепкодушный» панэтизм, рас-

сматривавший бытие как онтологизацию морального воззрения на мир, всё же уступил место «инновационному» пути развития. В последекабристской России началась «фаза надлома» (теория этногенеза Льва Гумилёва) русской целостности. При Пушкине «пришёл конец той "органической" России, памятник которой воздвиг в "Войне и мире" Толстой. Затем русский дух оказался загнанным в подполье или изгнанным на чужбину»- пишет академик РАН А.Панченко в предисловии к моей книге «Русское-от Загоскина до Шукшина» (Санкт-Петербург, 1992). Нравственность есть правда - лейтмотив возрожденческой «русопятой» прозы жизни – «деревенщинков». Голос князя Щербатова из XVIII века «О повреждении нравов в России» звучит серьёзным предупреждением современников о пагубности чуждого нации реформирования. Он справедливо полагал, что нравы допетровской Руси были здоровее и правильнее для «сохранения народа», и что лучше было бы вообще обойтись без реформ. Если бы воцерковленный духовный костяк народа и имперской державности не были крепки, то Россию не пришлось бы ломать через колено дважды за один XX век: сперва - большевиками-ленинцами, потом - антибольшевиками-ельцинцами. Ныне Россия нацелена на полное восстановление своей целостности и имперской самоидентификации.

* * *

14) Перманентный конфликт цивилизаций вступил в предвоенную конфронтацию. Россия, как определяет её нынешний судьбоносный момент английский историк А.Тойнби, сопротивляется имплантированию чужого «цивилизационного клина». И без закона Ньютона понятно, что сила противодействия русских теоретически должна быть не слабее силы воздействия на них. Однако тот факт, что страны Восточной Европы почти беспрепятственно со стороны РФ кооптированы в Североатлантический альянс, говорит о пассивизме русской внешней политики, об отсутствии стратегического паритета между антагонистами. Геополитическое пространство России стремительно сужается, как бальзаковская «шагреневая кожа». Запад вплотную приблизился к границам Российской Федерации. «Красная линия» Кремля сигнализировала об опасности.

15) Как справедливо полагает политолог С.Хантингтон, «линии разлома между цивилизациями - это есть линии будущих фронтов». Не потому ли в 1945-м американский генерал Джордж Паттон гнал без передышки своё воинство встречно с русскими – чтобы помешать маршалу Георгию Жукову занять тогда всю Европу.

Если тогда «линия межцивилизационного разлома» пролегла близ германского города Торгау на реке Эльба; то сейчас –считанные минуты ракетно-подлётного времени хоть от Румынии, хоть от Польши. А ежели Украина войдёт в НАТО – Запад приблизится к РФ до 5-7 «ракетных» минут подлёта до Москвы! Во избежание худшего сценария, Кремль 17 декабря 2021 потребовал письменной гарантии прекращения военного осваивания Украины и отведения всей военной инфраструктуры НАТО на позиции 1997-го. Иными словами: предложено капитулировать в ещё неначатой войне. В древнем трактате «Искусство войны» китайского полководца Сунь-Цзы сказано: «Сражения и захваты не могут считаться высшим мастерством военных действий, таким мастерством является сдача армии противника без боя».

16) Вряд ли коллективный Запад согласится на капитуляцию. Ведь «Россия очень сильно уступает НАТО как по людским и по промышленным ресурсам, так и по общей мощи вооружений...» - считает военный эксперта Б.Юлин. Разве что уравнивает наступательные силы сторон, так это наличие и у РФ гиперзвукового оружия и мощного ядерного арсенала. РФ – не СССР, угрожавший в случае форс-мажорных обстоятельств задействовать план академика А.Д.Сахарова: создать между Канадой и Мексикой «пролив имени Сталина». РФ – бизнеса ради заправляет танки «незалежной» российским горючим. И только когда всё «смайданилось» в гордиев узел, РФ «была вынуждена что-то делать» (слова В.Путина). Если бы не запахло жареным интересам правящего в РФ олигархата, вряд ли Кремль прибег бы к дерзкой риторике войны – предупреждению об “адекватных военно-технических мерах” в случае форс-мажорных обстоятельств.

* * *

17) России перед смертельным вызовом самому её суверенному существованию востребовалась официальная идеология сдерживания и реванша. «Мы хотим, чтобы в России была официальная, иже не прейдеши, идеология, базирующаяся на учении Русской православной церкви. Мы хотим, чтобы эта идеология была единственным или хотя бы главным основанием для внешней политики Государства Российского» - волеизъявление патриотического Царьграда. Аналитики Запада придумали и название этой идеологеме - «стратегический консерватизм».

18) Кремль только заигрывает с русским национализмом, но боится его (как и либеральный Запад) осуществления в полной мере. Русскому же народу потребно, чтоб Россия снялась с «нейтраль-

ного» дрейфа и двинулась от «общечеловеческой» демагогии к русским религиозно-корневым смыслам своего имперского существования. Поэтому актуализировалась выверка национальной идентичности России – фундаментала государства. Срочно реконструируется великодержавность России, основанная на её исторических традициях: идеях особого пути, самобытности, суверенности, русской цивилизационной особости (как сказал поэт: «у ней особенная стать: в Россию только можно верить»). Упорядочивается внутренний и международный ландшафт в соответствии с православно-советскими ценностями.

19) Благодаря выдвиженю собственного бренда «стратегический консерватизм» Кремль, по оценке экспертов вашингтонского Центра стратегических и международных исследований Х.Конли и Д.Руя в статье «The Kremlin Playbook 3» («Кремлёвские планы») стремится достичь следующих целей:

- Снизить прозападные настроения в целевых странах;
- Усилить поддержку политических действий России (внутри страны и за рубежом) и узаконить нарративы Кремля;
- Подорвать поддержку членства в ЕС среди государств-членов и уменьшить поддержку членства в ЕС и НАТО в странах-кандидатах;
- Сохранить страны постсоветского пространства в сфере влияния России;
- Подорвать внутреннюю сплочённость, суверенитет и, возможно, территориальную целостность таким образом, чтобы это поддерживало интересы Кремля (например, Босния);
- Сместить или ослабить руководство Вселенского Патриархата (который рассматривается как препятствующий объединению православного мира под руководством России); и
- Отменить санкции (сопутствующая и долгосрочная выгода) и подтолкнуть западные правительства к учёту политических интересов России.

20) Так, по оценке западных аналитиков, выглядят «стрелки на карте наступления» новой российской идеологии «стратегический консерватизмом» (Strategic Conservatism). Время покажет: «сосредотачивается» ли Россия для своего назревшего национального возрождения под своими древними путеводными Христовыми хоругвиями как «символе победы над смертью и

диаволом», а не использует всуе ради сиюминутной актуаловки (не дать Украине вступить в НАТО и спрогнозировать успешно для нынешнего престолонаследия результат президентских выборов-2024)?

21) Для России первостепенна задача восстановить свой исконный мировоззренчески геополитический «код», представляющий собой совокупность ключевых представлений россиян о своём месте в истории и мире, внешнеполитической стратегии и национальных приоритетах. Россияне пробуют как китайцы в политике: по всем договорам неизменно требуют «дуйдэн» — паритета взаимоотношений, мер и шагов. В соответствии с духовным концептом «инь и ян» («хаос и порядок»): упорядочивание расущностей — прекращение действия хаоса. В этой связи можно рассматривать требование к НАТО вернуться к состоянию до 1997-го – началу самодвижения альянса к расширению. И недопускать Западом военного освоения Украины – блюсти «красные линии» РФ. А для этого нужен России паритет сил с Западом, достижимый ассиметричным мегаоружием, способным в смысле ответных контругроз уничтожить США и Европу. И сопровождающая консервативно ценностная идеология Русской победы.

* * *

22) Подлинные консервативные ценности - как маховик локомотива — необходимы любой общественной системе, чтоб ультра-либеральный «майданизм» не разнёс убийственно саму государственную машину. Идёт дикое разжигание этнического национализма. Шовинистически раздувается ненавидеть к другому народу просто за то, что он есть (трамповцы, россияне). Запахло в США госпереворотом, даже Второй Гражданской войной. Идёт трансформация к закреплению однопартийной системы (со всеми средневеково-«-совковыми» вытекающими: «кто не с нами – тот против нас») – тенденция к лево-фашизирующему социализму. Политические лагеря – всё те же: левые (sinistram), значит греховные, плохие, и правые (iustum) - верные, правильные. Правые, будучи более консервативными, опирались на традиционную американскую этику: индивидуализм, самодостаточность, трудолюбие, независимость от государства, законопослушание, и равноправие перед законом. Левые с их марксистскими взглядами тяготели к социальным ценностям: коллективизм, стадность, подчинение начальству, примат государства над личностью, и экономическое равенство. Эти два подхода реализовались в двух партиях: республиканской и демократической. Пока Антифа и BLM под лозунгом «Долой полицию» без-

баррикадный плюрализм социумных ориентаций.

23) Язык политкорректности «новоязил» по Дж. Оруэллу: грабёж стал «перераспределением богатства», чёрный криминал стал «жертвой расизма,» талант стал «белой привилегией», «белый» стал означать «расист». «Нас, - как констатирует J.Fraden, - ждёт Вторая Гражданская война. Только она будет способна спасти страну. Это будет не только необходимо, но и законно. Напомню, что по этому поводу отцы-основатели США написали 244 года назад в Декларации Независимости: «Когда длинный ряд злоупотреблений и насилий, неизменно подчинённых одной и той же цели, свидетельствует о коварном замысле вынудить народ смириться с неограниченным деспотизмом, свержение такого правительства и создание новых гарантий безопасности на будущее становится правом и обязанностью народа». Консервативной реставрации «старых времён» жаждет и русское национальное сознание, дважды поломанное: в 1917 и 1991.

24) Всё возвращается на круги своя. Как утверждал евразиец Пётр Савицкий, «кто бы ни победил в Гражданской войне – "белые" или "красные", - всё равно Россия будет противостоять Западу, все равно она будет великой державой, все равно она создаст Великую Империю». Если так, то для созидания Евразийской России подходит неовизантийская модель государственности - основанная на сочетании религиозных ценностей Православия с началами Империи во главе с Самодержцем (как вариант – функцией его: Вождь, Генсек).

25) На Византизме «симфонии властей», где Церковь и монархия/«вождь народа» тесно сотрудничают в едином социальном литургическом делании - всеобщем спасении базируется идея «Пятой Империи» Александра Проханова. Чтоб воссияло Мономахово «государство правды» - торжество справедливости, спасения, добра. Истинный консерватизм может быть только верным хранением Божественной Истины, а не конъюнктурных задач правящих «элит». Государство и Церковь едины только если они выполняют общекорневую национальную миссию: вместе с народом осуществляют преображение себя и мира. Митрополит Илларион предсказывает русским великое духовное будущее, применяя к ним евангельскую истину «последние станут первыми».

26) Сквернословят по поводу сурового облика русского имперского консерватизма, уподобляя его «сатанизму сталинизма» и классифицируя «евразийские» проекты Путина и Дугина как

«правоэкстремистский интеллектуализ в неоавторитарной России». «Гиперконсерватизм» - вообще пугало для либералов. «В конечном счёте,- как полагает политолог А.Малашенко,- этот гипер ведёт к распаду государства – возможно, под напором возмущённой улицы, а возможно, и под ударами экстремизма». Историк А.Минаков же полагает, что консерватизм в России всегда коррелирует с сильной, централизованной, мощной иерархической властью. Ибо в русских условиях только такая власть может обеспечить необходимую мобилизацию как материальных средств, так и людских ресурсов для ведения многочисленных войн. Выжить можно только при наличии мощнейшей централизованной власти, воспринимаемой народом как благо и отождествляемой с тем, что сейчас называется цивилизационным кодом. Основополагающие смыслы духовной корневой системы жизнеобеспечения России – это неразрывно связанная разновидность триады Духовность, Державность и Соборность – содержание "Русской идеи."

27) Церковный раскол XVII века нарушил целостность сей Триады. Возник острый конфликт между светской и духовной властями, закончившийся утверждением первенства власти царя над властью патриарха. Старообрядности (двуперстию) неизменил протопоп Аввакум. Его рьяную приверженность традиции наследуют охранители/ почвенники/консерваторы – сторонники нереформируемого Православия.

28) Путинское понимание консерватизма - не адмирала А.С. Шишкова, одним из первых заговорившего о том, что вестернизированный верхний слой России превратился в некий особый народ, живущий в пределах большого народа, который сохранил подлинные русские ценности. При сохранении пропасти в РФ между «золотояхтовиками» и «подзаборным» людом не исключён смертный бой «двух Россий»: между исходными «проекциями» либералов/западников и консерваторов/хранителей «имперских амбиций». Это предчувствие сшибки я выразил в своей статье «Преодоление системных угроз национальной безопасности России»: «Говоря о национальной безопасности современной постбеловежской России, следует уточнить о какой из двух существующих ныне «Россий» идёт речь. Ведь либеральная свистопляска 1990-х годов поломала страну через колено на две неравные противоборствующие «России»: олигархически-мафиозную и подзаборно-нищенствующую. Отечество, духовно оккупированное и разграбленное несправедливой приватизацией, фактически расколото на

вотчину сверхбогатеев и жалкое бытийство обездоленного большинства. Разорван социум на два антагонистических лагеря. Соответственно, у каждого из этих образований - свои мировоззренческие и миросозерцательные критерии истины, разные потребности, несравнимые вызовы и угрозы, виды на будущее и сценарии стратегического развития страны».

29) И вот недавно В.Путин предложил свой проект новой государственной идеологии – «здоровый консерватизм». Социолог Карл Маннгейм считает, что понадобившаяся особая идеология может служить защите определённого политического или социального порядка от угрожающих ему внешних и внутренних вызовов. Вот этапы сотворения её.

* * *

30) Беловежским заговором был ликвидирован Советский Союз. Началась новейшая история России – сперва в общности СНГ, а затем в облике обкромсанной, десуверенной, коматозной РФ, воссиявшей щедринским восторгом глуповцев, освободившихся от себя самих (12 июня 1994 – день официального празднования «независимости России». От чего: разве она была колонией СССР?)! Отечество, раздербаненное в «лихие 90-е» «криминал-революционерами» («Великая криминальная революция» - термин кинорежиссёра С.С.Говорухина), лишили идеологии «совка» (Красной империи), запретив 13-й статьёй Конституции РФ установление взамен её какой-либо иной «в качестве государственной или обязательной». Рулящий РФ компрадор-олигархат вполне устраивал установившийся коллаборативно-корпоративный строй «экономики трубы», с политикой договорняка «газ со скидкой».

31) Но Божьим промыслом и путинскими усилиями госпереворот оказался недоворотом (моя формулировка). Россия возрождается из пепла и руин. В ситуации эскалации угроз и вызовов РФ актуализировался вопрос о создании общенациональной идеологии – гармонизанторе социума, оптимизаторе духа народа и «тротиле» информационно-организационного оружия в геополитическом противоборстве глобальных национальных интересов.

32) Правда, у класса начальников наверняка есть нечто по устранению угрозы «Берёзовый революции» в РФ. Это, по характеристике философом Ольги Малиновой, есть набор «сравнительно устойчивых и узнаваемых систем смыслов». Но сия методология скомпрометирована: «возникла коллизия бесправия одних при диктате других

членов общества», и стал очевиден потенциал 'злоупотребления властью' для ослабления конкурентных шансов оппонентов, вплоть до ограничения идеологического плюрализма путём запрета на высказывание тех или иных идей в публичных средах. ... Властвующая элита не вправе использовать государственные инструменты принуждения, чтобы навязывать собственные представления как обязательные или исключать право на высказывание иных точек зрения».

33) Верно, Госдума нередко принимает законы, с позиции непримиримой оппозиции, - сравнимые с «оккупационными». Многие формы общественного протеста объявлены незаконными. Взять, к примеру, штрафы или принудработы за «Оскорбление представителя власти» (УК РФ, Статья 319). Страшно далеки такие «законники» от народа. Россия, выбитая беловежским путчем из своей привычной жизни и даже цивилизационной колеи, всё никак не может полноценно самоидентифицироваться и упорядочиться в Законе. Творимый прожект капитализма с человеческим лицом не получается в смысле гармонии масс: слишком жаден до сверхприбылей олигархат, да при скупости властей на достойное исполнение «социальных гарантий». А проект СССР-2 – пока кошмарит воображение ГУЛАГом.

34) Относительно вопроса о взаимосвязанности плюрализма, госидеологии и личностной свободы... Либеральный консерватор правовед Борис Чичерин (право участвовать в государственной власти) рассматривал как высшее развитие личной свободы и её единственную гарантию: «Пока власть независима от граждан, права их не обеспечены от её произвола: в отношении к ней лицо является бесправным». Консервативно-либеральная концепция более органична для учёта всей полифонии спектра разный взглядов в социуме. Для гармонизации социума явно недостаточно рецептуры мультяшно-ельцинской демагогии кота Леопольда: «Ребята, давайте жить дружно!». Нет объявленного «консенсуса» обворованных и воров – бенефициарами приХватизации. Вопиющее социальное неравенство между «нищебродами» и «златояхтовиками» - никак не согласуется с тезисом жить в добром согласии «по справедливости и по закону».

35) Постсоветская идеосфера три десятилетия вырабатывала крайний антиэтатизм - минимизацию роли государства и пропагандировала вместо совести прогматику безбожного обогащения. Такой чудовищный крен парадигмы России был закреплён суперпрезидентской Конституцией. Потому подорванные силы минимизированного и ослаблен-

ного государства спровоцировали серию терактов с 1991 (Чечня и Дагестан) по 2004 (Беслан). Но Владимиру Путину удалось переломить тогда «пагубный мегатренд» (регресс, дерегуляцию) укреплением «вертикали власти» и ускорением национальной самоидентификации.

36) В октябре 1994 С.Шахрай и В.Никонов обнародовали «Консервативный манифест» — «консерватизма с российским лицом». В нем содержались основные постулаты классического консерватизма, подкреплённые цитатами от У.Черчилля до К.Леонтьева. В то же время предпринимается попытка создания идеологической доктрины «демократического патриотизма (В.Шумейко, В.Костиков). Составными элементами «новой идеологии» стала концепция формирования политической нации «россиян»: общепатриотическая риторика, включавшая в себя идею России как великого государства, заявления о необходимости реинтеграции постсоветского пространства, с ведущей ролью России как «первой среди равных», воскрешение формулы «единая и неделимая Россия». Однако теория «новой российской нации» не прижилась. Применение же формулы «единая и неделимая Россия» в «наведении конституционного порядка в Чеченской республике» положительно сработала. Российская государственность устояла перед сепаратистским вызовом «Ичкерии».

37) С 1996 по август 1998, по оценке историком Сергеем Пантелеевым, - активное время поиска «национальной идеи», призванной консолидировать общество. После непростой победы на президентских выборах («голосуй – или проиграешь!»), ещё более обостривших идейную поляризацию российского общества, Б.Ельцин 12 июля 1996 инициировал процесс разработки единой национальной доктрины, поручив выяснить «какая национальная идея, национальная идеология – самая главная для России» («Независимая газета». 1996, 13 июля, с.1). Однако кризис 17 августа 1998 (впервые в мировой истории государство объявило дефолт по внутреннему долгу) подвёл черту под проводившемся с 1992-го социально-экономическим, политическим и идеологическим курсом.

38) Этап с сентября 1998 по конец 1999 - возгонка «консервативной волны», на гребне которой осуществлялась операция «преемник» - транзит власти без потрясения/революции. Консервативная доктрина Е.Примакова -стратегия укрепления страны с построением «социально ориентированного рынка с государственным участием» - отвечала ожиданиям власти и общества. Идеологическая ниша «просве-

щённого» консерватизма (завуалированная триада графа С.С.Уварова) закрепилась программным положением о необходимости «национальной идеи», понимаемой как «патриотизм, державность, государственничество и социальная солидарность» («Независимая газета», 1999, 30 дек., с. 4). Но одно дело провозгласить консервативно-державный курс, а другое – реализовать. Как заявил тогдашний серый кардинал Кремля Владислав Сурков, «мы, конечно, безусловные консерваторы, хотя пока и не знаем, что это такое». Консерватизм как модный бренд. Д.Трамп побеждает Х.Клинтон (США), Н. Фарадж – Д.Кэмерона (Англия), движение «Пять звёзд» - брюссельскую бюрократию (Италия), В.Орбан - силы, олицетворяемые Д. Соросом (Венгрия). Один из номеров влиятельного журнала «Foreign Affairs» констатирует факт: «В наши дни никому в Америке, кажется, не хочется быть либералом - или хотя бы слыть им». А консервативное издание «National Review» печатает выдержку из книги философа Й.Хазони под заголовком «Либерализм как империализм». Уже в 2018-м другую книгу этого автора на ту же тему - «Достоинство Национализма» («The Virtue of Nationalism») многие объявили самой важной публикацией консервативной мысли со времён знаменитой книги Хантингтона «Столкновение цивилизаций».

* * *

39) Такова тенденция времени. И Россия – не исключение: идёт от западничества – к патриотизму, от радикализма – к консерватизму, от «свободного рынка» – к государственничеству, от идеологического нигилизма – к единой национальной идее. В.Путин называет свою идеологию то «здоровым», то «разумным», то «умеренным» консерватизмом. Известный российский идеолог и евразиец Александр Дугин в серии очерков «Либерализм - угроза человечеству» писал: «Интуитивно стремясь сохранить и восстановить суверенитет России, Путин вошёл в конфликт с либеральным Западом и его глобализационными планами, но и в альтернативную идеологию это не оформил». А разве он собирался «оформлять» какую-то развёрнутую антизападную идеологию? Вряд ли. Ибо его конфликт с либеральным Западом не имеет антагонистического характера. «В отличие от Бориса Ельцина В.Путин не радикал. В отличие от советских лидеров он не догматик».

40) Путинский консерватизм есть идеология сохранения общества таким как есть на данный момент (конечно же, с перспективой имманентного развития). Понятно: эволюционные перемены, «без потрясений». Но поскольку общество поляризовано – то обоим противопоставляемым нельзя

угодить. А вот выбор президента кому угодить – за ним. Его валдайская речь об идеологии базируется на «философии неравенства» философа Н.Бердяева. В ней не сказано о намерении организовать общественную жизнь по справедливости для всех и каждого члена общества. А значит «консервируется» какое есть: заведомо несправедливое устройство социума. И от народа трудно ожидать любви к такой власти.

41) А ведь учитывать бы надо и рефлекторную реакцию народа: нутряной консерватизм русского сознания. Это феномен притяжения к традиции, как этногравитационная константа (мой термин). Например, у Достоевского персонаж Фалалей – видит постоянно сон про белого бычка. Ещё пример – от политика В.С. Черномырдина: «Какую бы общественную организацию мы ни создавали - получается КПСС». Да и сам русский «консерватизм» разве функционально не таков же? Ведь в политической российской науке он синонимичен понятию «монархия», понимаемой как Самодержавие. Так что «охранительство» подлинных ценностей русской корневой системы априори должно быть проимперским, православно-народным – с приматом справедливости, совести, чести, доблести – Правды превыше всего. Такова потребность органичного существования русского народа.

42) Путинская идеологическая заявка не крайностна, а общегуманитарна, без кардинальных новаций: «Консервативный подход - не бездумное охранительство, не боязнь перемен и не игра на удержание, тем более не замыкание в собственной скорлупе. Это прежде всего опора на проверенную временем традицию (да, но на какую именно: имперскую, недавнюю советскую, или воровскую?.-м.р.Е.В.), сохранение и преумножение населения, реализм в оценке себя и других, точное выстраивание системы приоритетов, соотнесение необходимого и возможного, расчётливое формулирование цели, принципиальное неприятие экстремизма как способа действий».

43) Путинский идеологический концепт базируется на бердяевской «философии неравенства». Если это намёк на крепёж системы неравенства в российском социуме, то трудно отделаться от мысли о Божье-законном противодействии народа в ответ на такое злоупотребление властью. А бунт в России, по Пушкину, «решительный и беспощадный». Востребованная российским обществом неотложная тотальная справедливость не может дальше игнорироваться популистским режимом, допускающим лишь косметическое прихорашивание

его антинародной неприглядной сущности, а не ожидаемый демонтаж изначально чуждой и гнилой конструкции. Хотя Кремль заверяет, что возврата к 1937-му не будет, телекомментатор Вл.Соловьев то и дело предлагает власти назначить его карающим мечом органов -ГлавСМЕРШевцем. Всё не столь однозначно в этом «королевстве кривых зеркал».

44) А к Н.Бердяеву, если и апеллировать, – то с большей осторожностью. Ибо этот философ-экзистенциалист– «великий путаник», и может завести мимо цели. Более подходящим для выработки кремлёвской новой идеологии мне представляется зачинатель русской политологии профессор Московского университета Б.Н.Чичерин (пусть и объявил его Н.А.Бердяев «врагом демократии»). Б.Чичерин отстаивает конституалистский политико-правовой идеал. Руководящей идеей его «Курса государственной науки» было стремление «примирить начала свободы с началами власти и закона». Таков базовый постулат чичеринской программы охранительного либерализма, основной политический лозунг которого – «либеральные меры и сильная власть». К тому же, у него технологически умелее сопряжено «неравенство» со «справедливостью»: «Равенство состояний столь же мало вытекает из требований справедливости, как и равенство телесной силы, ума, красоты». Равенство прав (формальное) нельзя заменить равенством состояний (материальным). В русской политической культуре мало кто обосновывал столь убедительно онтологическую природу неравенства. Разве что ещё философ и дипломат Константин Леонтьев, для консерватизма которого Лев Толстой и Фёдор Достоевский - лишь «розовые христиане», не в должной мере препятствующие тенденции отпадения русского народа от Православия. Глубокая воцерковленность побудила его стать монахом.

45) «Президент Путин, - писал я в *2007*-м,- наверняка знаком с либерально-консервативной концепцией Бориса Чичерина, сформулировавшего политический принцип, весьма подходящий для нынешней власти в России: *"либеральные меры и сильная власть"*. "Это был, - характеризует Чичерина Бердяев, - редкий в России государственник, очень отличный в этом и от славянофилов, и от левых западников... Он принимает империю, но хочет, чтобы она была культурной и впитала в себя либеральные правовые элементы». Пункты, созвучные с путинскими.

46) И ещё штрих в пользу чичеринского лекала для кройки идеологии. Пупутал-таки Бердяев.

«Консерватизм, - пишет философ, - это не то, что мешает идти вверх и вперёд, а то, что мешает идти назад и вниз, к хаосу». По контексту, хаос присущ состоянию до консерватизма. У Путина же «хаос» - гипербола, употребление в переносном смысле, уподобление подрывной (террористической) хаотизации масс. Бердяевский импульс другого замеса: да, «напор хаотической тьмы снизу» (это верно), но не анархо-протестантов, а «перводродно-греховной зверино-хаотической стихии в человеческих обществах». Некие досоциуиные фантомы. А главное у Бердяева вот что: «хаотическая бесформенная тьма сама по себе не есть ещё зло, а лишь бездонный источник жизни». Путинская же мысль о бытовом «экстремизме» не вытекает из бердяевского онтологического «зла». Вообще с высшим советничеством по «злу» плохо обстоит дело в РФ, хотя Институт философии РАН потратил 742 тысячи рублей на исследование об аде и зле.

47) Зло не изначально, а постфактумно. Конфуций тому свидетельствует: «Зло не имеет самостоятельной причины в мироздании». Зло не могло быть создано добрым Небом в качестве самостоятельного элемента мира. Оно проистекает от нарушения порядка (не имеется в виду «общественный порядок»!-м.р.Е.В.), то есть от нарушения добра непониманием небесного порядка. Мы вносим в мир беспорядочность, разрушая изначальную гармонию, мы создаём в нем хаос, тем самым нарушая и уничтожая первоначальный порядок. Так появляются несчастья и беды, так появляется зло. Таким образом, оно есть результат нарушения мирового баланса или упорядоченности. Зло – это разбалансированность мироздания. По Бердяеву, хаос органично коррелирует с порядком. Он позитивная первокатегория мироздания (как в китайской «космогонии»), а не из разряда «массовых беспорядков» (дисгармония, бунтарство).

48) Так что самый раз индоктринировать в новую идеологию элементы либертарианского консерватизма: правое политическое течение, которое стремится объединить либертарианские и консервативные идеи, то есть, пытается развить идею сохранения традиций и сохранения консервативного курса развития, при этом сохраняя индивидуальную свободу отдельного человека. Соборность позволяла человеку видеть в себе личность или, говоря словами А. И. Клибанова, христианство воспитывало внутренний суверенитет личности. По абрису и сущности путинская идеологема – «либертарианский консерватизм», а потребность же России, как осаждённой крепости – иная: «стратегический кон-

серватизм Победы».

49) Консерватизм может быть компонентом в разной мировоззренческой комбинаторике. Так что «идеология порядка и охранительства» подвопросна. Ведь «порядок» в лозунге «анархия - мать порядка» - не легитимен для «охранительства». И то же самое – о власти. А если это «отмытое самовластье»? Консерватизм для России – естественный исторически и ментально выверенный консолидатор народа и предпосылочное условие установления нового баланса общественных интенций — между «иконой» (порядком) и «топором» (бунтом).

50) Не веря в добровольную смену «неизменного» курса страны, последователи народно-монархической стержневой русской ориентации уповают на спасение Отечества «национальной военной диктатурой». О путчевом сценарии смены власти пишут и на Западе: например, в книге Кэтрин Белтон «Люди Путина: как КГБ захватил Россию, а затем взялся за Запад» (Catherine Belton,"Putin's People: How the KGB Took Back Russia and Then Took on the West"). Ясно одно: без смены парадигмы нынешнего курса РФ - «национализации» самой логики мышления и усиления прорусской доминанты – страна утратит не только суверенитет, но и жизненные ресурсы для воспроизведения нации. Деградация и депопуляция народонаселения РФ – демографическая катастрофа с геостратегическими последствиями (с численно малым народонаселением в 145 975 300 человек–трудно сохранить Родину от Калининграда до Владивостока).

51) «Главным началом» путеводной идеологии должно стать усвоение аксиомы: Русская идея – имперская по сути – «зиждитель национального самосознания, культуры и религиозного промысла, как предназначение нации». Её категории Духовность, Державность и Соборность триедины - по образцу Святой Троицы и триады Добро, Истина и Красота. И помнить, что и либерал не жупел, и консерватор – не спасение, если не живут Россией. Как Герцен говорил о западниках и славянофилах: *они «смотрели в разные стороны», а «сердце билось одно»*. У таких своих задор молодости мудреет с возрастом, без сшибки друг с другом живут: «Кто в молодости не был либералом – у того нет сердца, кто в зрелости не стал консерватором – у того нет ума».

52) В плане конкретного примирения идеологических лагерей Русского мира вспоминается один биографический факт. В конце 1970-х мы с философом Петром Болдыревым первыми в Зарубежье попытались примирить

между собой «либералов» и «консерваторов». В совместной статье «Солженицын и Янов», появившейся в довлатовском «Новом американце» ("New American", New York), №59, 24-31 марта 1981, с.36), вместо привычного «либо-либо» мы объединили полюса, заменив эту расточительную бесплодную дизъюнкцию обнадёживающим «и-и». Задались вопросом: а не просматривается ли в исторических концепциях оппонентов, в их интерпретациях русской истории некоего «единства противоположностей» - при всём видимом и неоспоримом дуализме, очевидной несводимости их одна к другой? Оказалось, что и «прогрессисты» (либералы), и «интуитивисты» неизменных ценностей в истории (консерваторы - сторонники «народной субстанции», «общины», «народной души») идеологически «впадали» в русский «либеральный консерватизм» или «охранительный либерализм» Б.Чичерина, которого высоко ценили как либерал П.Струве, так и консерватор И.Ильин. Кредо консервативного (конкретного) либерализма можно выразить примерно так: не «либерализм вообще», а включающий национально-культурные традиции; не провинциальный, ущемлённо-национальный «консерватизм» (охранительство), а покоящийся на общечеловеческих культурных ценностях. Иногда в этих идеологических азимутах можно запутаться. Например: в Англии либерализму восемь веков, начавшемуся Великой Хартией вольностей - Magna Charta Libertatum. Стало быть, либеральные ценности свободы личности - более традиционные. Значит ли это, что приверженность именно ценностям либерализма оказывается самым махровым консерватизмом?

53) За поставленный нами тогда риторический вопрос: не пора ли современной русской оппозиции (с её историографией) обратить более пристальное внимание на этот третий, «средний» путь – дрейф конфронтаторов от противоборства к синтезу—нам досталось по полной от тех и других. Здоровый баланс консерватизма и либерализма коррекционно совершенствует обоих эти духовные устремлённости, нейтрализуя их крайние формы самовыражения — впадания консерватизма в «обскурантизм», а либерализма — «экстремальный прогрессивизм». Дефиниция между социальными группами сложнее, чем с аналогами из естествознания. Металлы «тоже» (как и идеологии) подвержены трансмутации и добрососедствуют с себе подобными - металлоидами. Классификация элементарна: неспаренные *d-электроны дают жизнь переходным металлам.* А тайны социума – приоткрываются дьявольскими отмычками: научили нейросеть

определять политические взгляды человека. Искусственный интеллект может по фотографии выявить сторонников либерализма и приверженцев консервативных взглядов. Оказывается: либералы чаще смотрят прямо в объектив фотокамеры, выражая удивление. У приверженцев же консервативных взглядов на лице просматривается отвращение. Либералы не переносят даже запаха консерваторов. Экспериментально вывели, что консерваторы брезгливее либералов. Вывод далеко идущий: возможно, то, что либералы практически не испытывают отвращения ни к чему, и толкает их на перемены. На протесты. Или даже на революционные преобразования. Зато консерваторы оказываются счастливее либералов: чем более либерален человек, тем он несчастнее. И наоборот. Объясняют феномен разным пониманием справедливости. Радоваться либералам мешает ощущение того, что разница между бедными и богатыми слишком велика, что общественные блага распределены как-то неправильно. От этого они, как правило, несчастливы даже в личной жизни.

* * *

54) Новый идеологический конгломерат консервативно-либерального наследия всего лучшего неизбежно аккомодирует в себе базовые ценности «классического» либерализма, такие как: абсолютная ценность человеческой личности и естественное («от рождения») равенство всех людей; существование определённых неотчуждаемых прав человека, таких как право на жизнь, личную свободу (в пределах чужой), справедливость; создание государства на основе общего консенсуса с целью сохранить и защитить естественные права человека; верховенство закона как инструмента социального контроля и «свобода в законе» как право и возможность «жить в соответствии с постоянным законом, общим для каждого в этом обществе и не быть зависимым от непостоянной, неопределённой, неизвестной самовластной воли другого человека» (Дж. Локк); способность каждого индивида к духовному прогрессу и моральному совершенствованию. «Участвуя в экономической жизни общества, - завещает Адам Смит из XVIIIвека, - каждый человек, кроме удовлетворения собственных интересов, поневоле способствует реализации общих интересов, ибо они - не что иное, как “сумма интересов отдельных членов общества”».

55) Основными чертами консерватизма считается: «Сохранение древних моральных традиций человечества; уважение к мудрости предков; неприятие радикальных изменений традиционных ценностей и институтов; убеждённость в том, что общество

нельзя построить в соответствии с умозрительно разработанными схемами; счастье невозможно без гармоничных отношений с обществом» (античные философы и консерватор Э.Берк). В современном консерватизме разные группировки объединены общими концепциями, идеями, идеалами. Хотя консерватизм традиционно отождествляется с защитой общественного статус-кво, характерной чертой современного консервативного ренессанса стал тот факт, что именно неоконсерваторы и «новые правые» выступили инициаторами изменений, направленных на перестройку существующего порядка. Так что не исключено вспенивание смуты как слева, так и справа. Между крайностными общего больше, нежели между центристами.

56) Видимо, такая двойственность закономерна. Ведь в либерализме изначально присутствует консервативный элемент. Так у мэтра либерализма Джона Локка, в его теории общественного договора значится неполная передача народом своих властных полномочий государству: они лишь делегируются для защиты всего общества. Иными словами, создавая государство, люди стремились обеспечить свои гражданские интересы, защитить свою свободу. «Речь, - как пишет исследователь вопроса Н.Андреев, - идёт о свободе совести, мысли, защите своего правового статуса, исполнении налоговых обязанностей, санкционированных государством и закреплённых в законодательстве и т.д. В силу изложенного либералы отстаивают необходимость принятия в каждом государстве Конституционного акта, закрепляющего права и свободы личности и гражданина». Славянофилам же, исповедующим внутреннюю правду как первостепенный регулятор общественных отношений, претило регламентированное законичество «договорняка». По их убеждению, Самодержавие и Народ единятся общим исповедываемым Православием. Как писал И.Аксаков: «в том-то вся и сущность союза Царя с народом, что божественная нравственная основа жизни у них едина, единый Бог, единый Судия, един Господень закон, единая правда, единая совесть». В основе деятельности и Земли (народа), и Царя являются Божественная воля и Правда. Следовательно, согласно славянофильской доктрине, ни о каких договорных началах возникновения власти речь идти не может. Монархия – лучшая для России форма правления, обеспечивавшая подлинную свободу.

57) Страшно далека от народа правящая номенклатура. Если Россию 1917-го погубили именно крайние «либералы» – почему такие до сих пор во власти?! В ситуации, когда народ и власть сосу-

ществуют как бы в параллельных реальностях, и «слуги народа» прислуживают правящему компрадор-олигархату, нередко принимая явно антинародные законопроекты, - большинство граждан РФ ждёт реальной пронародной политики, реализуемой не важно какой партийной эмблематики, лишь бы «своей в доску» по интересам и неподкупным принципам.

58) Всё что угрожает русской идентичности, не должно санкционироваться властями. Разве что в ограниченном разовом употреблении: как император Павел посоюзничал с еретикам и-католиками в противодействии Французской революции. А то с перестройкой фактически ликвидировали принцип соборности, и в качестве главной власти церкви был объявлен не собор, а администрация патриарха и архиереев. Митрополит Иларион после встречи с Папой Франциском сказал: «В православном народе существует очень большое предубеждение против католиков, и мы никоим образом не должны рисковать единством наших церквей и миром в наших церквях, поэтому мы должны во взаимоотношениях с католиками продвигаться с той скоростью, с которой возможно». Случайно ли, когда Россия сосредоточивается, Патриарх Кирилл предостерегает «начальников» от тирании? Или срабатывает ельцинская несменяемая система «сдержек и противовесов»? Тоже «традиция». Консерватизм - это верность изначальному Замыслу Божию, то есть православной религия и соответствующая идеология устройства земной государственной жизни. Это «цветущая сложность» цивилизационной специфики.

59) Попытки законсервировать правящий в РФ режим государственным «консерватизмом», «совершенно противоположным русской национальной традиции и осознанию русского народа как соборной личности», чрсвато потрясениями и деградацией страны. Для имперца монархического склада сознания неприемлем госконсерватизм, охранительный и от чрезмерных претензий Запада, и от собственного народа - узаконенный олигархической «многонациональной» конституцией и умножением карательных «экстремистских» статей в УК. К сожалению, и нынешнее церковное руководство нередко понимает свой консерватизм как жреческое служение любой власти, которой якобы не бывает «не от Бога». Но в нынешних условиях либерально-майданный бунт мог бы привести к власти в России правителей, ещё более враждебных русской традиции. «Идеология Победы как национальный проект» призвана противостоять идеологии «инклюзивный капитализм» нового глобалистского

порядка имени Клауса Шваба. Сам Божий Промысел обязывает Россию сыграть роль того самого Удерживающего, о котором говорил апостол Павел (2 Фес. 2: 7), в этой битве против мирового зла.

60) Сегодня российская власть позиционирует себя как приверженец консервативных ценностей, но действует чаще как прагматик и «дистиллированный либерал». Писатель Александр Проханов на вопрос, что плохого в либерализме, отвечает: «Опыт общения России с либералами кончается для страны совершенно трагично — Россия распадается. Либералы могут быть яркими и прекрасными, но их присутствие во власти приводит к тому, что они разрушают государство, не предлагая никаких других основ. Россия рушится и для её восстановления приходится возвращать имперскую форму правления огромной ценой — ценой потери исторического времени и потерей людей». Левые и правые отечественные консерваторы являются «державниками», для которых сильное государство является одной из ключевых ценностей. Россия в настоящее время все ещё борется с наследием «третьей смуты» 1991. Это – и потерянные территории, и экономические проблемы, и «негласная поддержка Вашингтонского консенсуса».

61) Французская исследовательница современного русского консерватизма Джульетт Фор пишет в своей диссертации о создающейся в идеологии на основе синтеза традиции и модерна, динамическом консерватизме (Juliette Faure.(2019) *L'idee de tradition au cœur de la politique du regime russe contemporain: un «conservatisme dynamique»?.*) Автор цитирует главного идеолога «динамического консерватизма» Виталия Аверьянова, утверждающего, что стоит задача «создать кентавра из православия и инноваций, из высокой духовности и высоких технологий. Этот кентавр будет представлять лицо России XXI века». Видимо, без «кентавра» - устрашающего симбиоза мудрости и смелости в лице древнего славянского божества Китовраса никак. Ибо, если верить Збигневау Бжезинскому, нельзя выиграть в холодной войне не имея альтернативной идеологии: «Чтобы быть военным противником США в мировом масштабе, России придётся выполнить какую-то миссию, осуществлять глобальную стратегию и...обрести идеологическую основу». Хотя Бжезинскому это кажется «маловероятным», Россия пробует разные варианты «спасения себя и мира». Прикидывает грядущий строй - меритократии (власть лучших) под эгидой специально созданного официального Стратегического совета РФ, оснащённого мощным «идеологическим оружием». Что однако не должно мешать, как го-

ворит философ И.Ильин, «душе русского народа всегда искать своих корней в Боге и в его земных явлениях: в правде, праведности и красоте».

62) Понятно, ныне греки, неовизантийство – «это всё наше». Но с позиции временной «обратной перспективы» (термин Д.С.Лихачёва), те «передние веки» для нас – самое важное: там первопричина нынешней «бесцелостности» РФ, как в ментальном, так и территориальном смыслах. Без строгой, ригористичной оценки случившегося с Родиной – от Никонова раскола до расстрела парламента – не выработать национальной идеологии русского народа.

63) С распадом Советского Союза была ликвидирована, как пережиток, и «совковая» идеология. Однако в условиях обострившегося ныне цивилизационного противоборства между историческими антагонистами российскому истеблишменту для консолидации нации потребовалась патриотическая идеология. Так родилась идеология «стратегического консерватизма». Апробирована она «второй зимней войной» - по «денацификации» Украины (термин на манер искоренения нацистской идеологии Entnazifizierung). Встаёт вопрос: если целью этой спецоперации провозглашено уничтожение «нацистского режима» бандеровцев, то для чего Кремль ведёт сепаратные переговоры с «украми», предварительно не решив сперва своей поставленной задачи?! Путин отступает, наторкнувшись на неожидаемое (просчёт разведки) сильное сопротивление противника? Или это тактика победы «иными средствами» (усыпить бдительность противника сменой кнута на пряник)? Ведь изначальная заявка на денацификацию широко контекстна: она подразумевает не только зачистку ЦЕЛИКОМ Украины, НО И ШИРЕ (Польша? Приднестровье? Прибалтика?) Что же заставило Путина пойти на «Хасавюрт2»? Только ли маячащий за проектом сепаратного договора интересы олигарха Абрамовича и иже с ним? Кремлю опасно терять патриотическую личину в глазах народа, ожидающего «взятия Киева» (чтобы жертвы были не напрасны). Вот и мечется правящая «элитита» между Сциллой и Харибдой — между двумя плохими вариантами исхода операции. Кремль в ситуации Zugzwang «принуждение к ходу» — положение в шахматах, в котором любой ход игрока ведёт к ухудшению его позиции.

Д-р Евгений Александрович Вертлиб/Dr. Eugene Alexander Vertlieb *президент Международного института стратегических оценок и управления конфликтами (МИСОУК-Франция); ответственный редактор отдела прогнозирования политики Запада «Славянской Европы» (Мюнхен); экзекьютив член Инициативы «Лиссабон-Владивосток» (Париж)*

BLOCKCHAIN: A Cyber Defense Force Multiplier

Dorian Belz

What is Blockchain?

t its core, a blockchain is simply a database for storing digital information where each *block* of time-stamped, published data is *chained* together with all preceding and subsequent blocks. What makes blockchain unique from traditional data storage is that data is not stored or controlled within a centralized database, but rather database management is shared by a decentralized network of trustless nodes that each maintain an identical copy of the blockchain. Before data can be published to the blockchain, it must first be validated by a majority of network nodes—typically referred to as *miners* or *validators*, depending on the consensus methodology used. This concept of "distributed consensus" makes the blockchain inherently tamper resistant and ensures that only verified transactions will be published to the blockchain.[1] The two primary consensus methodologies used in blockchain technology today are *proof-of-work*, first popularized in 2008 by the founder(s) of the Bitcoin[2] blockchain, and *proof-of-stake*, which has been used extensively in the development of derivative blockchain projects to help solve the issues of scalability and energy consumption associated with proof-of-workbased blockchains.

Both proof-of-work and proof-of-stake consensus protocols have associated strengths and weaknesses that must be considered when determining what type of blockchain to potentially use for Department of Defense applications. For proof-of-work blockchains, a considerable amount of CPU processing power is required to solve algorithmic puzzles before a block of data is published to the blockchain. The need for significant computational capacity and the security of distributed consensus make it fundamentally difficult for a dishonest node to manipulate data within a proof-of-work blockchain; however, the energy consumed by this protocol presents concerns from an environmental standpoint, as well as substantial challenges to scalability. By comparison, proof-of-stake blockchains do not require nodes to supply a large amount of computational power to validate transactions, but instead rely on a network of validators that hold a specified stake in the tokenized digital currency associated with the proof-of-stake blockchain. In this sense, proof-of-stake blockchains may be more vulnerable to

1 Samantha Ravich, "Leveraging Blockchain Technology to Protect the National Security Industrial Base," July 10, 2017, https://www.fdd.org/analysis/memos/2017/07/10/leveraging-block chain-technology-to-protect-the-national-security-industrial-base/

2 Satoshi Nakamoto, "Bitcoin: A Peer-to-Peer Electronic Cash System," 2008, https://bicoin.org/bitcoin.pdf

doi: 10.18278/gsis.7.1.8

data manipulation or denial-of-service attacks if bad actors are to accumulate a sufficient stake to serve as a trusted node on the network; however, methods such as "checkpointing"—where new transactions are verified against a master copy of the blockchain before being accepted onto a node's local reference version—can be used to rapidly identify and correct for these anomalies.[3]

Why is Blockchain Important to the DoD?

Our current defensive cyber architecture is rapidly becoming insufficient to establish and maintain battlespace superiority in the cyberspace domain of tomorrow. A report for Congress published in March 2020 by the Value Technology Foundation ,in collaboration with subject matter experts from the U.S. technology industry, stated this problem clearly as follows:

> The next generation of emerging technologies, which includes Artificial Intelligence, smart drones, robots, and additive manufacturing, will make the U.S. military even more dependent on digital technology. In this environment, the U.S. military has become critically dependent on secure, timely, accurate and trusted data. Yet, as data has grown in importance, cyber warfare has emerged to challenge the U.S. in the digital space. Today, key U.S. defense assets, ranging from communication systems to supply chains, can be disrupted by bad actors attempting to degrade U.S. capabilities.[4]

Mounting an effective cyber defense against potential bad actors is no longer as simple as installing firewalls and updating virus protection software. The exploitation methods used by our adversaries in the Cyberspace Domain are becoming increasingly sophisticated[5] and require novel solutions that redefine the landscape of the battlefield. Blockchain technology is the we need to accomplish this objective and has already proven to be invaluable across dozens of use cases for the U.S. Government and our industry partners. The Department of Defense, intelligence community, and our allies must act now to incorporate the latest advancements in blockchain technology into our existing cyber defense strategy and TTPs if we have any hope of establishing and maintaining strategic advantage in the Cyberspace Domain.

3 Rong Zhang and Wai Kin Chan, "Evaluation of Energy Consumption in Block-Chains with Proof of Work and Proof of Stake," , 2020, https://iopscience.iop.org/article/10.1088/1742-6596/1584/1/012023/pdf

4 Value Technology Foundation, "Potential Uses of Blockchain by the U.S. Department of Defense," March 2020, https://www.valuetechnology.org/_files/ugd/be4f79_17bfc382080f45edace0318ff0d70059.pdf

5 Victoria Adams, "Why Military Blockchain is Critical in the Age of Cyber Warfare," March 5, 2019, https://media.consensys.net/why-military-blockchain-is-critical-in-the-age-of-cyber-warfare-93bea0be7619

The scope of potential military applications for blockchain in defensive cyber operations is vastly expanding as the technology continues to mature. Use cases can be found throughout all facets of warfare ranging from protecting the integrity of military supply chain data to the prevention of adversary intrusion into our Joint All-Domain Command and Control (JADC2) infrastructure. This point is perhaps best illustrated by considering the damage a malicious actor can inflict by simply interrupting a data transmission from a command-and-control system to a weapons system, effectively preventing the weapon from responding to an incoming threat. Under our current system of centralized command-and-control, a single vulnerability may be sufficient to take a weapons system offline at a critical point in a commander's decision cycle.

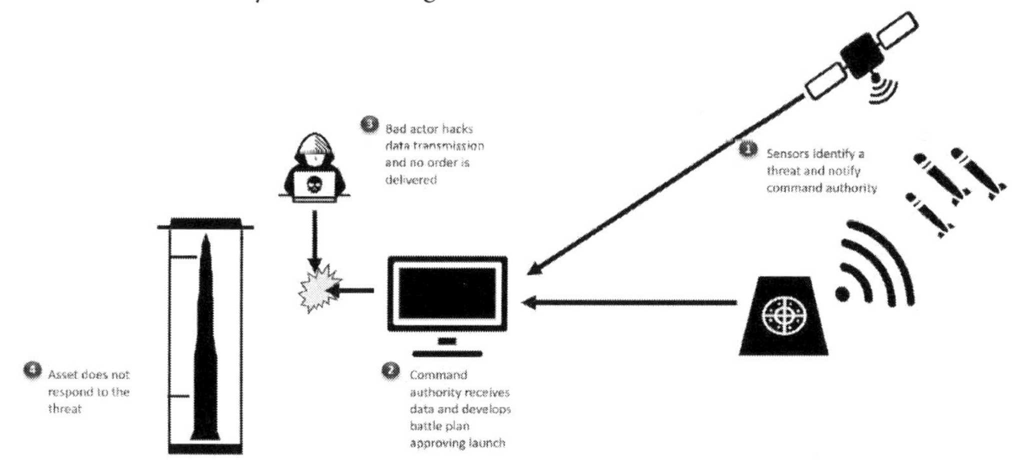

Now consider a similar scenario, but instead we substitute a blockchain-based, decentralized command-and-control infrastructure. In this case, the bad actor would need to not only access one computer system, but in essence the entire network of interconnected blockchain nodes simultaneously to interrupt the data transmission. In addition, any attempt to manipulate the data transmission or publish errant data to the blockchain would be immediately rejected by the honest nodes on the network through the chosen distributed consensus protocol.[6]

This hypothetical scenario is not entirely farfetched. Despite our best efforts to modernize and harden our command-and-control systems from a cybersecurity standpoint, many of our weapon systems have been in service for over a decade and continue to operate on outdated, legacy technologies. As bad actors continue to improve the complexity of their methods, it will become increasingly more difficult for cy-

6 Victoria Adams, "Why Military Blockchain is Critical in the Age of Cyber Warfare," March 5, 2019, https://media.consensys.net/why-military-blockchain-is-critical-in-the-age-of-cyber-warfare-93bea0be7619

ber defenders to protect our cyberspace terrain without the aid of advanced technologies like the blockchain.

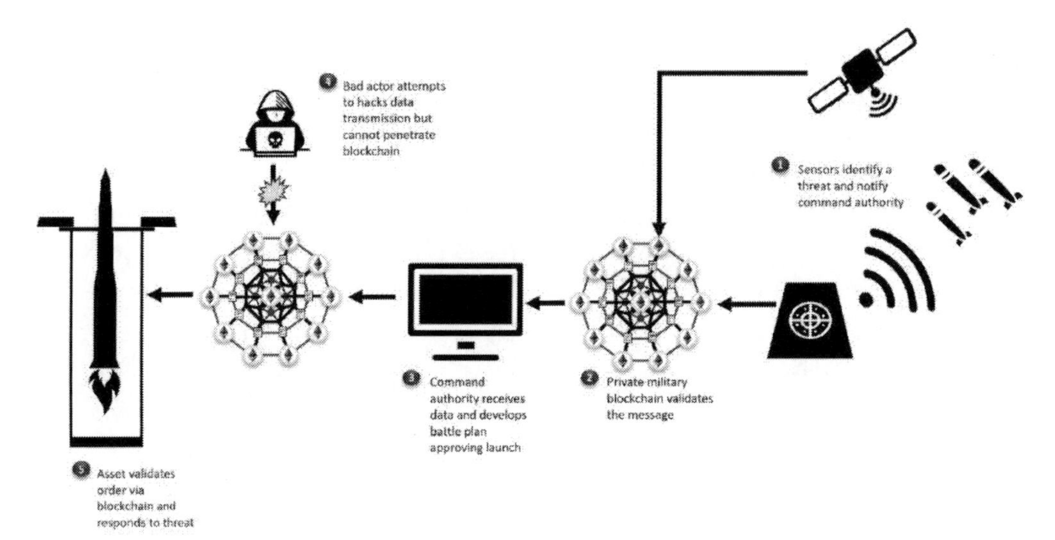

Why Must We Act Now?

In an era of Great Power Competition, the U.S. Department of Defense is quickly falling behind our near-peer competitors when it comes to incorporating blockchain technology at scale. As of January 2020, Chinese companies had filed over 10,000 blockchain related patents with the China National Intellectual Property Administration (CNIPA)—the Chinese equivalent of the U.S. Patent and Trademark Office (USPTO).[7] This equates to roughly 46% of global patent applications related to blockchain technology and is nearly double the number of patents filed in the United States.[8]

Additionally, in March 2020, the Chinese Government along with a consortium of Chinese banks and technology companies launched the Blockchain Service Network (BSN)—one of the first blockchains globally to be built and maintained by a central government.[9] Unlike the majority of early blockchains that operate publicly under a decentralized and permissionless system, access to China's BSN blockchain will be limited to only parties which have been granted appropriate permissions and transactions on the blockchain will not be publicly viewable to all participants. While a permissioned setup makes sense for an access-controlled, hardened classified network in use by a government organization like

7 Forkast News, "China is leading the global blockchain patent race," January 2020, https://forkast.news/china-is-leading-the-global-blockchain-patent-race/

8 Nestor Gilbert, "51 Critical Blockchain Statistics," January 2022, https://financesonline.com/blockchain-statistics/

9 Nick Stockton, "China Launches National Blockchain Network in 100 Cities," March 2020, https://spectrum.ieee.org/china-launches-national-blockchain-network-100-cities

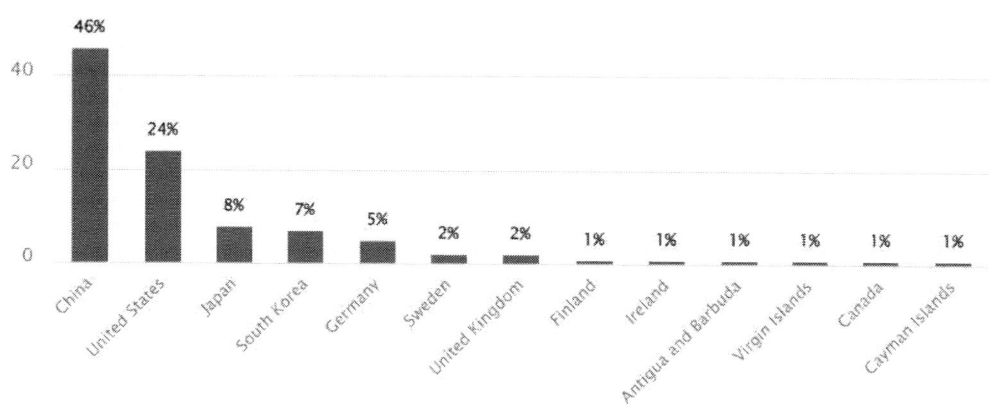

the U.S. Department of Defense, the BSN is instead marketed as a solution for broad commercial use cases. The lack of transparency over BSN blockchain transactions and overt control by the Chinese central government presents a significant risk that this medium will be used as a tool to collect on network participants, including potentially logistics and supply chain vendors of the U.S. defense industrial base.[10] It is no secret that President Xi Jinping intends to expand the BSN abroad, with a goal of creating a Chinese-controlled, blockchain-based, global infrastructure network—complete with its own central bank controlled cryptocurrency[11]—to give China a "new industrial advantage"[12] in the modern era. A delay by the U.S. Department of Defense to invest in the development and incubation of blockchain technologies will allow near-peer competitors like China to establish significant global market share and quickly propel them into a position as the partner of choice among countries in key strategic locations around the globe.

On May 13th, 2021, the People's Republic of China released their 14th Five-Year Plan for National Economic and Social Development, which for the first time indicates blockchain as a key accelerator for future economic development.[13] China's heavy invest-

10 Trevor Logan and Theo Lebryk, "America and its Military Need a Blockchain Strategy," April 2021, https://www.c4isrnet.com/opinion/2021/04/05/america-and-its-military-need-a-blockchain-strategy/

11 Alexander Zaitchik, Jeanhee Kim and Kelly Le, "Globalizing the Digital Yuan," June 2021, https://forkast.news/video-audio/part-ii-the-new-silk-road/

12 Miranda Wood, "Chinese President XI Jinping Emphasizes Importance of the Blockchain," October 2019, https://www.ledgerinsights.com/chinese-president-importance-of-blockchain/

13 Center for Security and Emerging Technologies, "Outline of the People's Republic of China

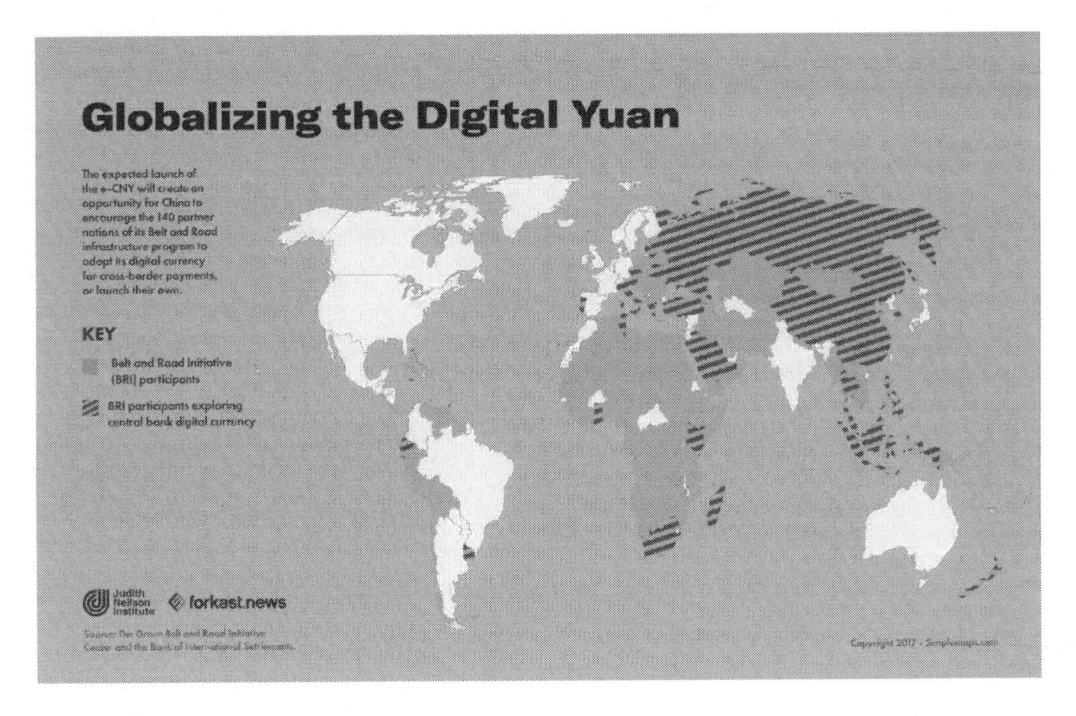

ments into blockchain technology over the past five years, and now stated intensions to continue investment for the next five, should serve as a warning to U.S. cybersecurity and defense officials. We are quickly yielding precious battlespace in the cyberspace domain of tomorrow with each passing day by delaying Department of Defense investment into developing blockchain technologies as a part of our defensive cyber infrastructure.

Recommendations & Conclusion

For the Department of Defense, the intelligence community, and our allies to maintain strategic technological advantage in the Cyber Domain, we must begin to invest our time and resources into developing blockchain-based technology solutions for military applications. To the maximum extent possible, we should examine and leverage best practices and lessons learned from other federal agencies, the defense industrial base, and our partners/allies in cases where blockchain technologies have already been implemented successfully. Additionally, in the traditions of ARPANET and SpaceX, we should continue to foster the Department of Defense relationships with public/private universities and technology companies to accelerate development of the blockchain-based cybersecurity infrastructure of the future.

14th Five-Year Plan for National Economic and Social Development," May 2021, https://cset.georgetown.edu/publication/china-14th-five-year-plan/

LCDR Dorian Belz holds a BS in Systems Engineering from the U.S. Naval Academy and is a graduate of the U.S. Naval War College and Joint C4I/Cyber Course at the Joint Forces Staff College. His 15-year naval career includes aircraft carrier deployments as an SH-60F helicopter pilot, headquarters assignments on the Korean Peninsula and at the Pentagon, and several years serving as an operations briefer for the Chief of Naval Operations and Secretary of the Navy. He transitioned to the Navy Reserves in early 2019 and began a civilian career as a technology consultant working for Ernst & Young, LLP. His professional focus areas include data management and analytics, data governance and quality controls, data privacy and security, and architecture design.

Intelligence Communities and the Media— The Case of the Danish Spymaster Lars Findsen

Ardavan M. Khoshnood

The relationship between the media and the intelligence community globally is vividly portraited in the book *Spinning Intelligence: Why Intelligence Needs the Media, why the Media Needs Intelligence*. At the beginning of their book, editors Professor Robert Dover and Professor Michael S. Goodman (2009) state that the relationship between "intelligence agencies, governments and the media" is "fluid, contradictory and occasionally supportive." The British Broadcasting Service's collaboration with the British Intelligence Service in disseminating anti-Soviet propaganda is one of many examples illuminating this complex relationship (Jenks, 2006). Media is often used by intelligence agencies to spread propaganda and information but is also used as an important source of open information. Although not all journalists are spies, intelligence officers often work undercover as journalists, allowing them to ask questions and be "noisy" without giving rise to suspicions (Braden, 1977). Indeed, media is also in need of intelligence agencies, for protection, but also for "leaks" from intelligence agencies which gives media material for reporting and discussions.

Many books and articles address the relationship between intelligence communities and the media. What has emerged is a complicated and complex relationship influenced by several different factors. While both have many similarities like collecting information and working with sensitive intelligence, they have two very different objectives. While the media in democratic countries seeks to expose information and contribute to public knowledge about different matters, intelligence communities globally strive to keep their intelligence, their *modus operandi*, and their intentions a secret. At the same time, reporters and journalists are highly vital to the intelligence community. In 1996, for instance, the Select Committee on Intelligence of the United States Senate had a briefing with respect to the Central Intelligence Agency's (CIA) use of journalists and the clergy in its operations (Select Committee on Intelligence of the United States Senate, 1996). One of the individuals questioned at the briefing, then CIA director John M. Deutch, stated: "I, like all of my predecessors for the last 19 years, have arrived at the conclusion that the Agency should not be prohibited from considering the use of American journalists or clergy in exceptional circumstances" (Select Committee on Intelligence of the United States Senate, 1996).

While the relationship between the media and intelligence communities in democracies is at least to some degree established and based on mu-

doi: 10.18278/gsis.7.1.9

tual respect, valuing the media's desire for openness and the intelligence communities need for privacy and secrecy, challenging conflicts do occur. One such conflict is currently on display in one of the world's most free and democratic countries—Denmark. During the writing of this paper in January 2022, the former head of the Danish Military Intelligence, Lars Findsen, remains in Danish police custody accused of leaking information to the media. Since this incident, Danish media has published several troublesome reports about the country's intelligence community. The Danish Security and Intelligence Service, at the same time, is trying to censor Danish media from publishing more news deemed to be of harm for Denmark's national security by the intelligence organization.

Intelligence Organizations and Media

Although it was during the 1960s the media (newspapers, radio as well as television) began to significantly cover various intelligence matters, it is believed that already during the Second World War, the media had showed interest in exposing intelligence communities and state secrecies (Moran, 2011). Although this relationship is highly important, it is not without complications.

One aspect in the media-intelligence relationship is how media is used by the intelligence community as a target of manipulation—that is, how intelligence organizations deliberately forward information to the media for disclosure and publication. This information may be false or true. What characterizes them, however, are the intentions of the intelligence organization, which are to manipulate the media for, in the eyes of the intelligence organization, a superior purpose (Bakir, 2017). For this objective, intelligence organizations either "leak" information to media, or use journalists and reporters connected to the intelligence community or even in secret employed by them (Magen, 2015). It is important to point out, however, that the use of the media by an intelligence organization is not always about manipulating the media. Using the media may be one way for the intelligence agency to show openness and transparency. The fact that most intelligence organizations today have a media department or a media liaison should be understood not only as a mean to show openness, but also for intelligence organizations to have access to the media. The intelligence communities' contact with media, and how transparent they are, is thus highly calculated (Teirila, 2016). Shpiro (2001), for example, in discussing German intelligence, the *Bundesnachrichtendienst* (BND), describes their relationship with the media as "defensive openness," meaning that "a limited amount of openness is maintained toward the media in order to influence media content."

The other aspect of the media-intelligence relationship, is media's role as a watchdog, thus investigating and reporting on the state and not least, its intelligence community (Bakir, 2017; Teirila, 2016). While the intelligence community tries to maintain control over the media, neither confirming or denying events; they also regularly try to censor the media (Bakir, 2017; Moran, 2011). In Israel, for instance, the Israeli Military Censorship (IMC) has extensive legal powers to not only shut down media, but also order materials in the media to be deleted. All Israeli media are also obligated to submit material discussing matters related to national security to the IMC before publication (Shpiro, 2001). This is, of course, much different in more democratic states. In countries like Norway, Denmark, and Sweden, for example, there are no such laws. The freedom of press is very strong in these countries. The media-intelligence relationship is thus highly related to the form of government.

While being a watchdog is one of media's democratic obligations (Shpiro, 2001), it is not without risk. In 1984, two terrorists were killed by Israeli security personnel after they had initially been arrested. A newspaper later revealed this atrocity, and that a secret inquiry was ongoing. For reporting this, the newspaper was shut down for several days (Magen, 2015). In Germany, the BND has through the years tried to stop publication of several articles and books for being critical to them (Shpiro, 2001). In the U.S., different administrations have taken tough positions towards "leaks" to the media with respect to intelligence and security matters (Hillebrand, 2012). During the 1990s, for instance, the CIA tried to humiliate and destroy the credibility of a journalist for revealing that the CIA, in cooperation with the Contras, had been bringing cocaine into the U.S. (Bakir, 2017). Media outlets, as well as individual journalists, risk being labelled "enemy of the state" when reporting on the intelligence community. Media's important role in overseeing the intelligence community is though clearly shown in an article by Loch K. Johnson (2014). He examined 10 intelligence failures and scandals in the U.S. and showed that high media coverage of an event also contributed to high oversight by Congress. Even though this important role of the media is acknowledged, most examples of collusions between the media and intelligence communities are from before the 1990s. This current piece adds to the current knowledge of the media-intelligence relationship with an example from Denmark in 2022.

The Danish Intelligence Community

There are currently two major national intelligence organizations active in Denmark, the Danish Security and Intelligence Service (*Politiets Efterretningstjeneste* (PET)) and the Danish Defence Intelligence Service (*Forsvarets Efterretningstjeneste* (FE)).

Prior to the Second World War, the Danish Security Police (SIPO) was established as part of the Danish Police. SIPO was dissolved in 1947 and replaced by the Intelligence Department of the National Police Chief (REA), which had been established in 1945. Six years later, in 1951, the PET was established (PET, 2022a). The main objective of the PET is to counter and prevent "threats to freedom, democracy and security in Danish society." The three main threats to the Danish national security, according to PET, are terrorism, political extremism, and espionage. Several different institutions conduct supervision and oversight of PET, among them the Ministry of Justice, the Parliament, and the Danish Intelligence Oversight Board (TET) (PET, 2022b). As of June 1, 2015, the head of the PET is Finn Borch Andersen.

During the Second World War there were two intelligence divisions in the Danish Military: the Intelligence Section of the General Staff, which was established in 1911, and the Intelligence Section of the Naval Staff. In 1950, these two sections were combined, creating the Intelligence Department. In 1967, the FE was established (West, 2008, 2015). The FE is Denmark's military intelligence and security, as well as its foreign intelligence organization. It is divided into six departments, and its main objective is to "prevent and counter threats against Denmark and Danish interests" (FE, 2022a). The Danish signal intelligence as well as cyber security and cyber operations, are also part of the FE (2002b). The oversight of FE is foremost conducted by the TET, but just like PET, several other institutions also share oversight responsibilities for the FE. The current head of the FE is Svend Larsen.

The Case of Lars Findsen

Lars Findsen, born in 1964, was the head of the PET between 2002 and 2007 before relocating to the Ministry of Defence. In 2015, he was appointed to the head of the FE, a position he held until August 2020 when he was suspended. The Danish Intelligence Oversight Board claimed that the FE had not only had withheld vital information from the board, but also provided them with incorrect information (TET, 2020). In December 2021, a commission that investigated the criticism forwarded by the TET acquitted both the FE as well as Findsen (Krog, 2021).

On December 9, 2021, the PET issued a press release stating that four members of the Danish intelligence community had been arrested the day before for leaking information. The press release stated: "They have been charged with violation of Section 109(1) of the Danish Criminal Code by having imparted highly classified information from PET and DDIS" (PET, 2021). DDIS is the acronym for the Danish Defence Intelligence Service, which is the same as the FE.

The four arrestees were later detained and their identities, according to Danish law, withheld from the public. On January 10, 2022, the court allowed the names to be published, and one of arrestees was Lars Findsen (Ryrsö et al., 2022). The media reported that Findsen had been under surveillance for a long time before being arrested (Fastrup et al., 2022). Exactly what Findsen is accused of is unknown, but he is being detained for violating §109 of the Danish Criminal Code, which states that an individual who discloses information related to "secret negotiations, deliberations or resolutions of the state or its rights in relation to foreign states, or which has reference to substantial economic interests of a public nature in relation to foreign countries" can be imprisoned for 12 years (Legislation Online, 2005). The law is thus about leaking information and not espionage, and the last time the law was used to convict someone was in 1980.

How, then, did the head of the FE, Lars Findsen, find himself in this chaos? Danish media, which has reported extensively on the matter, has revealed that the PET started to investigate Findsen and other members of the Danish intelligence community after the media reported on several highly sensitive and secret matters related to national security. It was after the criticism of the TET that the media started their investigation, and very soon, article after article was published revealing top secret information. During the summer of 2020, several news reports were published stating that some of the criticism was about the collaboration between the FE and the National Security Agency (NSA) of the U.S., with respect to the NSA's use of Danish fibre cables to spy on different subjects and individuals (Fastrup et al., 2020). In September 2020, the Danish newspaper *Berlingske* exposed details about the top-secret collaboration between the FE and the NSA (Bjørnager et al., 2020). Although all this had certainly cause frustration among the Danish intelligence community, the revelations continued. In March 2021, Ekstrabladet revealed that the FE had warned the Danish Government that children of ISIS-terrorists from Denmark, who are being held in captivity in Syria, could be kidnapped and trained by ISIS to conduct terrorism in their European countries (Khaja et al., 2021). In May 2021, the Danish Broadcasting Corporation reported that FE had collaborated with the NSA to spy on, among others, then German chancellor Angela Merkel (Fastrup et al., 2021).

All the above revelations made the Danish intelligence community highly uneasy. A massive investigation was conducted, and the head of the FE, Lars Findsen, was arrested and detained. He remains in custody awaiting prosecution. While the ordeal could have ended here, the Danish intelligence community, the PET and the FE, turned to the media. In a meeting with high officials from the Danish Broadcasting Corporation, Berlingske as well as JP/Politikens Hus, the FE, and the PET informed them that both journalists as well as their editors would be arrested and face up to 12 years imprisonment, if they disclosed any further information that threatened national security (Fastrup et al., 2022).

Discussion

The case of Lars Findsen and the threat to media by the intelligence community for not exposing what is deemed to be top-secret information of vital interest for the notion of national security, creates several problems that must be addressed.

Once again, the intelligence community and media go *tete-a-tete* in one of the world's foremost democracies. As far as we know, no reporter or editor of any media publications have been arrested in Denmark. They have, however, as explained above, been threatened with legal consequences and thus been censored. Danish media has reported that at least one article has been withdrawn by the Danish Broadcasting Corporation (Fastrup et al., 2022). Just like during the Cold War and prior decades, the intelligence communities' strongest tool in order to control the information flow is the law. There are, however, some disturbing questions which arise from this Danish case. Should it always be illegal in democracies to publish and disclose secret or top-secret intelligence and information related to the intelligence community? Does this apply to all kind of intelligence and information? What if the information is related to radical measures like torture or extra judicial killings? And if there is a consensus that reporters should be gagged with reference to matters related to national security, should they always remain silent, no matter what? Even though many papers have been written on the media-intelligence relationship over the years, many questions remain to be answered or at least discussed.

Focusing on the Danish intelligence community, the case of Lars Findsen is, to say the least, quite humiliating for Denmark. Its foremost spymaster has been arrested and detailed for months. Either Lars Findsen is a scapegoat and being used by the intelligence community to set an example, or Findsen—as one of the main intelligence and security figures of the country—has been leaking information and intelligence to media, thus contributing to destabilize and demoralize the Danish intelligence community. Whichever of the two options are true, one thing is for sure—Denmark's reputation among the world's intelligence communities has been seriously harmed. Can Denmark be trusted with vital information? Can collaborations with Danish intelligence proceed without new revelations—like the U.S. spying on Angela Merkel? And what consequences will Danish foreign and security policy face if these revelations continue?

Even though I have more questions than answers in my short discussion, there is one thing I am very sure about—in the current instable security atmosphere we are witnessing globally, this will be one of many times we witness a countries' intelligence community in open conflict with the media. The main questions, though, are how these conflicts will affect our transparency, openness, and democracy, and how these events will be exploited by authoritarian regimes.

References

Bakir, Vian. 2017. "News media and the intelligence community." In: *Routledge Handbook of Media, Conflict and Security*. Piers Robinson, Philip Seib and Romy Fröhlich (eds.). New York: Routledge.

Bjørnager, Jens Anton, Jens Beck Nielsen, Henrik Jensen, Steffen Nyboe Mcghie, and Simon Andersen. 2020. Et pengeskab på Kastellet har i årtier gemt på et dybt fortroligt dokument. Nu er hemmeligheden brudt. Berlingske. https://www.berlingske.dk/samfund/et-pengeskab-paa-kastellet-har-i-aartier-gemt-paa-et-dybt-fortroligt

Braden, Tom. 1977. Worldwide Propaganda Network Built by the C.I.A. *The New York Times*. https://www.nytimes.com/1977/12/26/archives/worldwide-propaganda-network-built-by-the-cia-a-worldwide-network.html

Dover, Robert and Michael S. Goodman. 2009. *Spinning intelligence: why intelligence needs the media, why the media needs intelligence*. Oxford: Oxford University Press.

Fastrup, Niels, Henrik Moltke, Trine Maria Ilsøe, and Lisbeth Quass. 2020. FE-skandale omhandler tophemmeligt spionagesamarbejde med USA. Danmarks Radio. https://www.dr.dk/nyheder/indland/fe-skandale-omhandler-tophemmeligt-spionagesamarbejde-med-usa

Fastrup, Niels, and Lisbeth Quass. 2021. Forsvarets Efterretningstjeneste lod USA spionere mod Angela Merkel, franske, norske og svenske toppolitikere gennem danske internetkabler. Danmarks Radio. https://www.dr.dk/nyheder/indland/forsvarets-efterretningstjeneste-lod-usa-spionere-mod-angela-merkel-franske-norske

Fastrup, Niels, Trine Maria Ilsøe, Lisbeth Quass, and Louise Dalsgaard. 2022. Hemmelig PET-taskforce aflyttede spionchef Lars Findsen i månedsvis for at afsløre læk til medierne. Danmarks Radio. https://www.dr.dk/nyheder/indland/hemmelig-pet-taskforce-aflyttede-spionchef-lars-findsen-i-maanedsvis-afsloere-laek

FE. 2022a. Hovedopgaver. https://www.fe-ddis.dk/da/arbejdsomrade-a/

FE. 2022b. Organisation. https://www.fe-ddis.dk/da/om-os/organisation/

Hillebrand, Claudia. 2012. The Role of News Media in Intelligence Oversight. *Intelligence and National Security*, *27*(5): 689–706.

Jenks, John. 2006. *British Propaganda and News Media in the Cold War*. Edinburgh: Edinburgh University Press.

Johnson, Loch K. 2014. Intelligence shocks, media coverage, and congressional accountability, 1947–2012. *Journal of Intelligence History*, *13*(1): 1–21.

Khaja, Nagieb, Jeppe Findalen, Magnus Mio, and Thomas Foght. 2021. Regeringen advaret af FE: Islamisk Stat smugler børn ud fra fangelejrene. Ekstrabladet. https://ekstrabladet.dk/nyheder/politik/danskpolitik/regeringen-advaret-af-fe-islamisk-stat-smugler-boern-ud-fra-fangelejrene/8502825

Krog, Andreas. 2021. Kommission frikender efterretningstjeneste og chefer. Altinget. https://www.altinget.dk/artikel/kommission-renser-efterretningstjeneste-og-chefer

Legislation Online. 2005. The Criminal Code. https://www.legislationline.org/download/id/6372/file/Denmark_Criminal_Code_am2005_en.pdf

Magen, Clila. 2015. Media Strategies and Manipulations of Intelligence Services: The Case of Israel. *International Journal of Press/Politics*, *20*(2): 247–265.

Moran, Christopher. 2011. "Intelligence and the Media: The Press, Government Secrecy and the 'Buster' Crabb Affair." *Intelligence and National Security*, *26*(5): 676–700.

PET. 2021. Four members of staff arrested for leaking information from the intelligence services. https://pet.dk/Nyheder/2021/Fire%20personer%20anholdt%20for%20laekager%20fra%20efterretningstjenesterne.aspx

PET. 2022a. Historie. https://pet.dk/Om%20PET/Historie.aspx

PET. 2022b. About PET. https://pet.dk/English/About%20PET.aspx

Ryrsö, Mikael, and Thomas Foght. 2022. Spionchef sidder fængslet i lækagesag. Ekstrabladet. https://ekstrabladet.dk/krimi/spionchef-sidder-faengslet-i-laekage-sag/9075714?ilc=c

Select Committee on Intelligence of the United States Senate. 1996. CIA's Use of Journalists and Clergy in Intelligence Operations. U.S. Government Printing Office: Washington, D.C.

Shpiro, Shlomo. 2001. The Media Strategies of Intelligence Services. *International Journal of Intelligence and CounterIntelligence*, *14*(4): 485–502.

Teirila, Olli. 2016. Intelligence and Media: Multidimensional Effects of Publicity. *American Intelligence Journal*, *33*(2): 137–143.

TET. 2020. Tilsynet med Efterretningstjenesterne afslutter særlig undersøgelse af Forsvarets Efterretningstjeneste (FE) på baggrund af materiale indleveret af én eller flere whistleblowere. https://www.tet.dk/wp-content/uploads/2020/08/PRESSEMEDDELELSE.pdf

West, Nigel. 2008. *Historical Dictionary of World War II Intelligence*. Plymouth: The Scarecrow Press Inc.

West, Nigel. 2015. *Historical Dictionary of International Intelligence*. New York: Rowman & Littlefield.

GLOBAL SECURITY AND INTELLIGENCE STUDIES JOURNAL

Call for Papers on

Strategic Deterrence

Subject: Call for Papers for *Global Security and Intelligence Studies* Special Issue

Global Security and Intelligence Studies (GSIS) is currently accepting submissions for its upcoming special edition on *Strategic Deterrence* (anticipated Summer 2023). The state of global power and politics have changed dramatically over the past few decades. Shifts in power projection capabilities and doctrine combined with revisionist regional powers, failing states, and global resource competition has shed new emphasis on the concept of deterrence amongst great powers.

The defining factors of *Strategic Deterrence* are vast and can include a number of variables such as nuclear, cyber, Economic, proxy warfare, evolution of tactics and doctrine, non-state actors, etc. *GSIS* welcomes original empirical research, critical analysis research, research notes, action notes from the field, policy relevant essays, and book reviews for the special issue. Book reviews should focus on the title's contribution to, or discussion of *Strategic Deterrence*.

Instructions for authors and submission procedures can be on the Journal's homepage: https://gsis.scholasticahq.com/for-authors

Please make all submissions by: **01 January 2023**

Submissions and inquiries can be made via the GSIS submission page, above, or by emailing the editorial team directly: Dr. Carter Matherly (Carter.Matherly@gmail.com) and Dr. Matthew Loux (matthew.loux67@mycampus.apus.edu).

Carter Matherly, PhD
Co-Editor in Chief

Matthew Loux, PhD
Co-Editor in Chief

20 June 2022

Related Titles from Westphalia Press

The Zelensky Method
by Grant Farred

Locating Russian's war within a global context, The Zelensky Method is unsparing in its critique of those nations, who have refused to condemn Russia's invasion and are doing everything they can to prevent economic sanctions from being imposed on the Kremlin.

China & Europe: The Turning Point
by David Baverez

In creating five fictitious conversations between Xi Jinping and five European experts, David Baverez, who lives and works in Hong Kong, offers up a totally new vision of the relationship between China and Europe.

Masonic Myths and Legends
by Pierre Mollier

Freemasonry is one of the few organizations whose teaching method is still based on symbols. It presents these symbols by inserting them into legends that are told to its members in initiation ceremonies. But its history itself has also given rise to a whole mythology.

Resistance: Reflections on Survival, Hope and Love
Poetry by William Morris, Photography by Jackie Malden

Resistance is a book of poems with photographs or a book of photographs with poems depending on your perspective. The book is comprised of three sections titled respectively: On Survival, On Hope, and On Love.

Bunker Diplomacy: An Arab-American in the U.S. Foreign Service
by Nabeel Khoury

After twenty-five years in the Foreign Service, Dr. Nabeel A. Khoury retired from the U.S. Department of State in 2013 with the rank of Minister Counselor. In his last overseas posting, Khoury served as deputy chief of mission at the U.S. embassy in Yemen (2004-2007).

Managing Challenges for the Flint Water Crisis
Edited by Toyna E. Thornton, Andrew D. Williams, Katherine M. Simon, Jennifer F. Sklarew

This edited volume examines several public management and intergovernmental failures, with particular attention on social, political, and financial impacts. Understanding disaster meaning, even causality, is essential to the problem-solving process.

Donald J. Trump, The 45th U.S. Presidency and Beyond International Perspectives
Editors: John Dixon and Max J. Skidmore

The reality is that throughout Trump's presidency, there was a clearly perceptible decline of his—and America's—global standing, which accelerated as an upshot of his mishandling of both the Corvid-19 pandemic and his 2020 presidential election loss.

Brought to Light: The Mysterious George Washington Masonic Cave
by Jason Williams, MD

The George Washington Masonic Cave near Charles Town, West Virginia, contains a signature carving of George Washington dated 1748. Although this inscription appears authentic, it has yet to be verified by historical accounts or scientific inquiry.

Abortion and Informed Common Sense
by Max J. Skidmore

The controversy over a woman's "right to choose," as opposed to the numerous "rights" that abortion opponents decide should be assumed to exist for "unborn children," has always struck me as incomplete. Two missing elements of the argument seems obvious, yet they remain almost completely overlooked.

The Athenian Year Primer: Attic Time-Reckoning and the Julian Calendar by Christopher Planeaux

The ability to translate ancient Athenian calendar references into precise Julian-Gregorian dates will not only assist Ancient Historians and Classicists to date numerous historical events with much greater accuracy but also aid epigraphists in the restorations of numerous Attic inscriptions.

The Politics of Fiscal Responsibility: A Comparative Perspective by Tonya E. Thornton and F. Stevens Redburn

Fiscal policy challenges following the Great Recession forced members of the Organisation for Economic Co-operation and Development (OECD) to implement a set of economic policies to manage public debt.

Growing Inequality: Bridging Complex Systems, Population Health, and Health Disparities Editors: George A. Kaplan, Ana V. Diez Roux, Carl P. Simon, and Sandro Galea

Why is America's health is poorer than the health of other wealthy countries and why health inequities persist despite our efforts? In this book, researchers report on groundbreaking insights to simulate how these determinants come together to produce levels of population health and disparities and test new solutions.

Issues in Maritime Cyber Security Edited by Dr. Joe DiRenzo III, Dr. Nicole K. Drumhiller, and Dr. Fred S. Roberts

The complexity of making MTS safe from cyber attack is daunting and the need for all stakeholders in both government (at all levels) and private industry to be involved in cyber security is more significant than ever as the use of the MTS continues to grow.

A Radical In The East by S. Brent Morris, PhD

The papers presented here represent over twenty-five years of publications by S. Brent Morris. They explore his many questions about Freemasonry, usually dealing with origins of the Craft. A complex organization with a lengthy pedigree like Freemasonry has many basic foundational questions waiting to be answered, and that's what this book does: answers questions.

This publication is available open access at:
https://gsis.scholasticahq.com/
http://www.ipsonet.org/publications/open-access

Thanks to the generosity of the American Public University System

This issue can be read online by visiting:

Made in the USA
Middletown, DE
24 September 2022

10745931R00128